AKADEMIEKONFERENZEN

Band 14

Proceedings of New Perspectives in Quantum Statistics and Correlations

Edited by
MORITZ HILLER · FERNANDO DE MELO
PETER PICKL · THOMAS WELLENS
SANDRO WIMBERGER

im Auftrag der Heidelberger Akademie
der Wissenschaften,
Akademie der Wissenschaften
des Landes Baden-Württemberg

Universitätsverlag
WINTER
Heidelberg

Bibliografische Information der Deutschen Nationalbibliothek

Die Deutsche Nationalbibliothek verzeichnet diese Publikation
in der Deutschen Nationalbibliografie;
detaillierte bibliografische Daten sind im Internet
über *http://dnb.d-nb.de* abrufbar.

ISBN 978-3-8253-6001-6

© 2012 Universitätsverlag Winter GmbH Heidelberg
Imprimé en Allemagne · Printed in Germany
Druck: Memminger MedienCentrum, 87700 Memmingen

Gedruckt auf umweltfreundlichem, chlorfrei gebleichtem
und alterungsbeständigem Papier

Den Verlag erreichen Sie im Internet unter:
www.winter-verlag.de

Preface

In March 2010, a group of about fifty scientists, including theoretical and experimental physicists as well as mathematicians, met in the ancient buildings of the Heidelberg Academy of Sciences and Humanities. Their common aim was to exchange views and knowledge on different aspects of quantum correlations and their statistical description. The challenge posed was big: Create bridges, communicate! The atmosphere was different from other conferences; this time, the other participants were not their usual fellows and some may have felt as if this had been their first conference. For several younger colleagues this was actually the case.

Long talks, of forty-five minutes each, facilitated the exchange of ideas. Presumably "simple" concepts, well known in their particular sub-field, had to be put into a language common to the entire audience. A renown mathematician's presentation, for example, could be followed by a talk of a young and highly motivated experimentalist. Mindsets changing over and over. The arising questions and comments would be of the most varied nature and form. The struggle for communication would continue during coffee-breaks, poster sessions, and even during social events. Day by day, the initial awkwardness was turning into vivid discussions.

All this was set in the beautiful scenery of Heidelberg with its long-standing tradition in science. Excursions to the city and the castle as well as the conference dinner completed the program and helped to establish links between the researchers of the different fields.

We are convinced that the concept of a mixed audience, both in disciplines and in experience, worked out very well and that the different groups profited quite a bit from the meeting. While the one side was astonished by the possibilities of state-of-the-art experiments and curious which physical effects still needed theoretical investigation, the other side was eager to see which of their discoveries was already understood theoretically or even with mathematical rigor. These proceedings witness the diversity of thoughts during the conference. We hope to have contributed to strengthen the communication among the various branches of physics and mathematics since a collective effort is indispensable to elucidate the nature of complex quantum systems.

This conference would not have been possible without the commitment of the Heidelberg Academy of Sciences and Humanities. Beyond the Academy for their generous financial support, we would like to thank all the people involved in the excellent organization, foremost Heidemarie Herburger and Dr. Herbert von Bose. Finally, we are indebted to the authors who contributed to the present proceedings as well as to Wolfgang Ketterle, the mentor of the conference.

Introduction

Some decades after Schrödingers famous 1926 papers "Quantisierung als Eigenwertproblem I-IV", the role of entanglement in quantum mechanical systems is being discussed more vividly than ever. New experiments and the prospect of applications like quantum computing and quantum cryptography have drawn the attention of many scientists to that area.

In contrast to classical wave equations, where, for example, a huge collection of water molecules is described by a single wave in three-dimensional space, in quantum mechanics already a single particle is described by a wave. In a system of many particles, the collection of all particles is then described by a wave in an abstract high-dimensional space. This description in terms of a collective wave is not just a formal difference between classical wave mechanics and quantum mechanics, but also gives rise to correlations in quantum mechanical many-body systems that are unknown in classical mechanics.

"When two systems, of which we know the states by their respective representation, enter into a temporary physical interaction due to known forces between them and when after a time of mutual influence the systems separate again, then they can no longer be described as before, viz., by endowing each of them with a representative of its own. I would not call that one but rather the characteristic trait of quantum mechanics, the one that enforces its entire departure from classical lines of thought. By the interaction the two representatives [the quantum states] have become entangled." (E. Schrödinger, "Discussion of Probability Relations between Separated Systems" Proceedings of the Cambridge Philosophical Society 31 (1935) pp. 555-562).

During the last decades, the theoretical understanding of single quantum systems, as well as the experimental ability to isolate, manipulate and control them, has reached an unprecedented level. Current efforts are focused on the next step, namely to attack few-body and in particular many-body systems. Depending on the statistics of the underlying particles — distinguishable or indistinguishable particles with Bose respectively Fermi statistics — different forms of correlations emerge. In many cases it is not possible or at least impractical to find the full solution of the Schrödinger equation. Therefore a better understanding of entanglement on an effective level is inevitable.

In the case of many-particle systems, the formation of Bose-Einstein condensates is an immediate consequence of the indistinguishability of bosons. At sufficiently low temperatures, a macroscopic number of atoms ($\sim 10^5$) in a harmonic trap become locked together in the lowest quantum state of the system. As the de Broglie wavelength of the particles increases, the single-particle identity is lost, and the whole behaves as a macroscopic matter wave, with a well-defined phase. This coherence can be used, for example, in order to create macroscopic entangled states between two modes inside a double well potential. On the other hand, turning on inter-atomic interactions, correlations between the particles emerge, which also may have a significant impact on the physical behavior of such a system.

Another experimental breakthrough, which allows to investigate the importance of quantum correlations for the description of many-particle systems, is the realization of optical lattices for ultracold atoms. When a Bose-Einstein condensate is loaded in a shallow lattice, the wavefunction of each atom is spread all over the lattice, inducing long-range correlations between distant lattice sites. By increasing the optical lattice potential and thereby the interatomic interactions, these correlations are destroyed, and the system turns from a superfluid to an insulating phase. Hence, by enhancing or diminishing the quantum correlations, completely different states of matter are created.

Although the extreme cases of this quantum phase transition are well understood, the same is not true for lattices of intermediate height. In this regime, the statistical properties of energy levels reveal correlations between them, and characterize the system as a chaotic one. Due to the enormous complexity of the related eigenstates, an efficient simulation of the system's complete dynamics on a classical computer is not possible for larger numbers of particles and lattice sites and a different, statistical, approach is needed.

Notwithstanding, simulation of quantum systems is one of the founding problems of quantum information science. Entanglement is believed to be the fuel for the promised speed-up of quantum computers when compared to classical ones. Nevertheless, the full power of this new paradigm is only captured in the limit of many-particle registers, i.e., only when multipartite entanglement can be created and coherently controlled in such a way that it prevails over the impact of unavoidable noise. Since entanglement measures are nonlinear functions of the quantum state, once more, their evaluation is a hard mathematical problem. A statistical treatment which determines the evolution of entanglement of typical states under noisy environments, and thus allows the construction of efficient control schemes, is highly desired.

The variety of the above mentioned examples shows that phenomena of quantum statistics and correlations are ubiquitous and represent a link between seemingly disjoint branches of physics and mathematics. While various attempts to establish statistical theories that capture correlations have been made within the individual subfields, the necessity for a broad quantum statistical approach is evident. To create a common ground for all the different aspects necessary to describe systems composed of many subsystems, as reflected by the following collection of articles, was the triggering idea that culminated in the conference NEW PERSPECTIVES ON QUANTUM CORRELATIONS AND STATISTICS.

Celsus Bouri, Moritz Hiller, Fernando de Melo,
Florian Mintert, Peter Pickl,
Thomas Wellens, Sandro Wimberger.

Contents

Effective evolution equations from many body quantum dynamics

Benjamin Schlein

Institute for Applied Mathematics, University of Bonn
Endenicher Allee 60, 53115 Bonn, Germany
benjamin.schlein@hcm.uni-bonn.de

Abstract. In these notes we review some recent results concerning the derivation of effective equations from first principle quantum dynamics. In particular, we discuss the derivation of the semi-relativistic Hartree equation for the evolution of boson stars, and the derivation of the Gross-Pitaevskii equation for the dynamics of Bose-Einstein condensates.

1 Introduction

Systems of interest in physics are typically composed by a huge number of elementary particles. Dilute samples of Bose-Einstein condensates contain $10^3 - 10^6$ atoms (and, strictly speaking, each atom contains many elementary components). The number of molecules contained in chemical samples is typically of the order of Avogadro's number, $N_A \simeq 6 \cdot 10^{23}$. Systems of relevance in astronomy and cosmology (like stars and galaxies) are composed by up to 10^{60} elementary components.

In principle, the dynamics of these systems can be determined by solving fundamental evolution equations like the Newton equation or the many-body Schrödinger equation. In practice, however, fundamental equations are impossible to solve (neither analytically nor numerically) when so many particles are involved (unless the interaction among the particles is neglected). Moreover, observers are not interested in determining the precise evolution of every particle. Instead, they need a prediction for the macroscopically measurable properties of the dynamics (which result by averaging over the many particles in the system). For this reason, it is very important to find effective evolution equations which, on the one hand, can be easily solved (numerically), and, on the other hand, accurately predict the macroscopic behavior of the system under consideration. One of the main goal of statistical mechanics consists therefore in the development of effective theories approximating the solutions of fundamental evolution equations in the relevant regimes. In these notes, we are going to discuss two examples of systems of interest in physics, for which the derivation of effective evolution equations can be made rigorous in a mathematical sense. In Section 2, we will illustrate the derivation of a semi-relativistic Hartree equation for the evolution of boson stars. In Section 3, we will sketch the derivation of the Gross-Pitaevskii equation for the dynamics of initially trapped Bose-Einstein condensates. In both cases, the starting point of our analysis is the fundamental many-body Schrödinger equation for bosonic systems. In the rest of the introduction, I will show in an abstract setting how effective evolution equations emerge from many body quantum dynamics in certain regimes.

We consider quantum mechanical systems of N spinless bosons in three dimensions (the spin does not play any role, and therefore will be neglected). We describe these systems on the Hilbert space $\mathcal{H}_N = L_s^2(\mathbb{R}^{3N})$, consisting of all functions in $L^2(\mathbb{R}^{3N})$ which are symmetric with respect to arbitrary permutations of the N particles. The time evolution is then described by the N particle Schrödinger equation

$$i\partial_t \psi_{N,t} = H_N \psi_{N,t} \tag{1}$$

for the wave function $\psi_{N,t} \in \mathcal{H}_N$. We consider Hamilton operators of the form

$$H_N = \sum_{j=1}^{N} -\Delta_{x_j} + \lambda \sum_{i<j}^{N} V(x_i - x_j), \tag{2}$$

where $\lambda \in \mathbb{R}$ is a coupling constant and where the potential $V(x)$ describes the (two-body) interaction (the precise form of V depends on the system under consideration). Here and in the following, we choose units so that Planck's constant $\hbar = 1$ and the mass of the particles $m = 1/2$. We restrict our attention to (approximately) factorized initial data, where all particles are essentially described by the same orbital. The kinetic energy is then of the order N, while the potential energy is of the order λN^2. To obtain a nontrivial effective evolution equation in the limit of large N, we have to assume that $\lambda N^2 \simeq N$, hence that $\kappa := \lambda N$ is a quantity of order one. This regime is known as the mean-field regime.

To analyze the mean-field regime, consider, at time $t = 0$, the factorized initial data $\psi_{N,t=0} = \varphi^{\otimes N}$, for an arbitrary one-particle orbital $\varphi \in L^2(\mathbb{R}^3)$. Because of the interaction, factorization is not preserved by the time evolution. However, if $N \gg 1$ and $\kappa = N\lambda$ is of order one, the interaction is very weak and one may still expect factorization to be approximately (and in an appropriate sense) preserved:

$$\psi_{N,t} \simeq \prod_{j=1}^{N} \varphi_t(x_j) \tag{3}$$

for an appropriate evolved one-particle orbital $\varphi_t \in L^2(\mathbb{R}^3)$. If this is true, it is very easy to derive the self-consistent nonlinear Hartree equation

$$i\partial_t \varphi_t = -\Delta\varphi_t + \kappa(V * |\varphi_t|^2)\varphi_t \qquad (4)$$

for the one-particle orbital φ_t. This simple argument suggests that the Hartree equation (4) gives an effective description of the evolution of initially factorized bosonic systems in the mean-field regime characterized by $N \ll 1$ and fixed $\kappa := N\lambda$.

To obtain a mathematical precise statement, we need to specify in which sense (3) holds true. To this end, we define, for $k = 1, \dots, N$, the k-particle reduced density matrix associated with $\psi_{N,t}$ by taking the partial trace

$$\gamma_{N,t}^{(k)} = \mathrm{Tr}_{k+1,\dots,N} \, |\psi_{N,t}\rangle\langle\psi_{N,t}|$$

where $|\psi_{N,t}\rangle\langle\psi_{N,t}|$ denotes the orthogonal projection onto $\psi_{N,t}$. In other words, $\gamma_{N,t}^{(k)}$ is defined as a non-negative trace-class operator on $L^2(\mathbb{R}^{3k})$ with kernel given by

$$\gamma_{N,t}^{(k)}(\mathbf{x}_k; \mathbf{x}_k') = \int d\mathbf{x}_{N-k} \, \psi_{N,t}(\mathbf{x}_k, \mathbf{x}_{N-k})\overline{\psi}_{N,t}(\mathbf{x}_k', \mathbf{x}_{N-k}) \,,$$

where $\mathbf{x}_k = (x_1, \dots, x_k)$, $\mathbf{x}_k' = (x_1', \dots, x_k')$, $\mathbf{x}_{N-k} = (x_{k+1}, \dots, x_N)$. Note the normalization $\mathrm{Tr}\, \gamma_{N,t}^{(1)} = 1$.

The next theorem, which holds under suitable assumptions on the potential V, tells us that (3) can be understood as convergence (in the limit of large N) of the reduced densities associated to $\psi_{N,t}$.

Theorem 1. *Let $\varphi \in H^1(\mathbb{R}^3)$, $\|\varphi\| = 1$, $\kappa \in \mathbb{R}$ and let $\psi_{N,t} = e^{-iH_N t}\varphi^{\otimes N}$ be the solution of the Schrödinger equation (1), with initial data $\psi_N = \varphi^{\otimes N}$ and with Hamilton operator*

$$H_N = \sum_{j=1}^{N} -\Delta_{x_j} + \frac{\kappa}{N} \sum_{i<j}^{N} V(x_i - x_j) \,. \qquad (5)$$

Then, if $\gamma_{N,t}^{(k)}$ denotes the k-particle reduced density associated with $\psi_{N,t}$, we have, for every fixed $k \in \mathbb{N}$ and $t \in \mathbb{R}$,

$$\gamma_{N,t}^{(k)} \to |\varphi_t\rangle\langle\varphi_t|^{\otimes k}$$

as $N \to \infty$. Here, the convergence is in the trace norm and φ_t is the solution of the Hartree equation (4) with $\varphi_{t=0} = \varphi$.

The first proof of Theorem 1 has been obtained by Spohn in [19] for bounded potentials. In [8], Erdős and Yau proved Theorem 1 for the (attractive or repulsive) Coulomb potential $V(x) = \pm 1/|x|$. In [18], a joint work with I. Rodnianski, we considered again the Coulomb interaction, but this time we obtained bounds on the rate of the convergence. In [11], Knowles and Pickl extend the theorem to more singular potentials (with control of the rate of convergence).

2 Dynamics of boson stars

In this section we consider systems of gravitating bosons forming so called boson stars. We describe boson stars with the Hamilton operator

$$H_{\text{grav}} = \sum_{j=1}^{N} \sqrt{1 - \Delta_{x_j}} - G \sum_{i<j}^{N} \frac{1}{|x_i - x_j|} \tag{6}$$

acting on the Hilbert space $\mathcal{H}_N = L_s^2(\mathbb{R}^{3N})$ (we use a relativistic dispersion for the particles, but the interaction is classical). As explained in Section 1, we are interested in the dynamics generated by (6) for large N and small G, with NG of order one. Since the physical value of the gravitational constant, in our units, is approximatively given by $G_{\text{phys}} \simeq 10^{-40}$ (for bosons with mass comparable to a hydrogen atom), this model can be used to describe boson stars with $N \simeq 10^{40}$ particles. For such values of N, it makes sense to fix $\lambda := NG$ and to study the dynamics generated by

$$H_N = \sum_{j=1}^{N} \sqrt{1 - \Delta_{x_j}} - \frac{\lambda}{N} \sum_{i<j}^{N} \frac{1}{|x_i - x_j|} \tag{7}$$

in the limit $N \to \infty$.

The Hamiltonian (7) is critical in the following sense. For every $N \in \mathbb{N}$, there exists a critical coupling constant $\lambda_{\text{crit}}(N)$ such that H_N is bounded below for all $\lambda \leq \lambda_{\text{crit}}(N)$ and such that

$$\inf_{\psi \in L^2(\mathbb{R}^{3N})} \frac{\langle \psi, H_N \psi \rangle}{\|\psi\|^2} = -\infty$$

for all $\lambda > \lambda_{\text{crit}}(N)$. It was proven in [15] that, as $N \to \infty$, $\lambda_{\text{crit}}(N) \to \lambda_{\text{crit}}^H$, where λ_{crit}^H is the critical constant for the Hartree energy

$$\mathcal{E}_{\text{Hartree}}(\varphi) = \int dx \left| (1 - \Delta)^{1/4} \varphi(x) \right|^2 - \frac{\lambda}{2} \int dx dy \frac{|\varphi(x)|^2 |\varphi(y)|^2}{|x - y|} . \tag{8}$$

In other words, λ_{crit}^H is such that $\mathcal{E}_{\text{Hartree}}(\varphi) \geq 0$ for all $\varphi \in H^{1/2}(\mathbb{R}^3)$ if $\lambda \leq \lambda_{\text{crit}}^H$ while, if $\lambda > \lambda_{\text{crit}}^H$,

$$\inf_{\varphi \in H^{1/2}(\mathbb{R}^3), \|\varphi\|=1} \mathcal{E}_{\text{Hartree}}(\varphi) = -\infty .$$

The criticality of the Hamiltonian (7) is a sign for the instability of boson stars when $\lambda = NG > \lambda_{\text{crit}}^H$. If the number of bosons in the star exceeds the critical value $N_{\text{crit}} = \lambda_{\text{crit}}^H / G$, the star collapses.

Next, we focus on the properties of the evolution generated by the Hamiltonian H_N defined in (7). For the subcritical regime, we show in [3], a joint work with A. Elgart, that the many body dynamics can be approximated (in the sense of Theorem 1) by the solution of the relativistic Hartree equation

$$i\partial_t \varphi_t = \sqrt{1 - \Delta}\, \varphi_t - \lambda \left(\frac{1}{|.|} * |\varphi_t|^2 \right) \varphi_t \,. \tag{9}$$

Note that, in the subcritical regime, Lenzmann showed in [12] that Eq. (9) is globally well-posed in the energy space $H^{1/2}(\mathbb{R}^3)$. What about the supercritical regime? Since H_N is not bounded from below, it is not a priori clear how to define the one-parameter group of unitary transformations $U_N(t) = e^{-iH_N t}$ describing the time-evolution. To circumvent this problem, we introduce a tiny, N-dependent, cutoff $\alpha(N)$ in the Coulomb potential,

$$H_N^\alpha = \sum_{j=1}^N \sqrt{1 - \Delta_{x_j}} - \frac{\lambda}{N} \sum_{i<j}^N \frac{1}{|x_i - x_j| + \alpha(N)} \tag{10}$$

and we assume that $\alpha(N) \to 0$ as $N \to \infty$. The regularized Hamiltonian H_N^α is now bounded below for every N. Therefore it can be extended (uniquely) to a self-adjoint operator on $\mathcal{H}_N$, the unitary group $U_N(t) = e^{-iH_N^\alpha t}$ is well defined and the Schrödinger equation

$$i\partial_t \psi_{N,t} = H_N^\alpha \psi_{N,t} \tag{11}$$

is globally well posed on $\mathcal{H}_N$. On the other hand, since the cutoff vanishes in the limit $N \to \infty$, we may still expect the effective dynamics to be described by the semi-relativistic Hartree equation (9).

It is important to notice that the criticality of the model can also be observed at the level of (9). For $\lambda > \lambda_{\text{crit}}^{\text{H}}$, the equation is still locally well-posed in $H^{1/2}(\mathbb{R}^3)$ (for an arbitrary initial data φ in $H^{1/2}(\mathbb{R}^3)$ there exists a unique solution φ_t in $H^{1/2}$ on the time interval $t \in (-T, T)$, for some $T > 0$). In general, however, the local solution cannot be extended to a global solution (i.e., one cannot take $T = \infty$). In fact, for $\lambda > \lambda_{\text{crit}}^{\text{H}}$, it was proven by Fröhlich and Lenzmann in [9] that there exist solutions of (9) which exhibit blow-up in finite time. This means that there are solutions φ_t of (9), and $0 < T < \infty$ such that

$$\|\varphi_t\|_{H^{1/2}} = \left(\int dx \left| (1 - \Delta)^{1/4} \varphi_t(x) \right|^2 \right)^{1/2} \to \infty \tag{12}$$

as $t \to T^-$. Solutions of (9) exhibiting blow-up in finite time can be used to give a dynamical description of the phenomenon of gravitational collapse.

The next two theorems from [16], a joint work with A. Michelangeli, show that, also in the supercritical regime, the solution φ_t to the relativistic Hartree equation (9) continues to approximate the many body dynamics until the time where φ_t blows up (if φ_t does not exhibit blow up, then it stay close to the solution of the many-body Schrödinger equation on every finite time interval).

The first theorem proves that, if φ_t does not blow up in the time interval $[-T, T]$, $|\varphi_t\rangle\langle\varphi_t|$ is close to $\gamma_{N,t}^{(1)}$, as $N \to \infty$, for all $|t| < T$.

Theorem 2. *Fix $\varphi \in H^2(\mathbb{R}^3)$ with $\|\varphi\| = 1$ and $T > 0$ such that*

$$\kappa := \sup_{|t|\leq T} \|\varphi_t\|_{H^{1/2}} < \infty \tag{13}$$

where φ_t is the solution of (9) with initial data $\varphi_{t=0} = \varphi$. Let $\psi_{N,t} = e^{-iH_N^\alpha t}\varphi^{\otimes N}$. Then there exists a constant C (depending only on T, $\|\varphi\|_{H^2}$, and κ) such that

$$Tr\left|\gamma_{N,t}^{(1)} - |\varphi_t\rangle\langle\varphi_t|\right| \leq \frac{C}{\sqrt{N}} \tag{14}$$

for all $|t| \leq T$.

The next theorem shows that if φ_t blows up at time T, then also the solution to the many body Schrödinger equation $\psi_{N,t}$ collapses as t approaches T, if, at the same time, $N \to \infty$ sufficiently fast. This result justifies the use of the Hartree equation (4) for the description of the gravitational collapse of boson stars.

Theorem 3. *Fix $\varphi \in H^2(\mathbb{R}^3)$ with $\|\varphi\| = 1$. Suppose that $T_c > 0$ is the first time of blow-up for the solution φ_t of (4) with initial data φ ($\|\varphi_t\|_{H^{1/2}} < \infty$ for all $t \in [0, T_c)$ and $\|\varphi_t\|_{H^{1/2}} \to \infty$ as $t \to T_c^-$). Let $\psi_{N,t} = e^{-iH_N^\alpha t}\varphi^{\otimes N}$, and assume that, in (10), $\alpha(N) \geq N^{-\beta}$ for some $\beta > 0$. Then, for every $|t| < T_c$, there exists a constant $C_t < \infty$ such that*

$$\|(1 - \Delta_{x_1})^{1/4}\,\psi_{N,t}\| \leq C_t$$

uniformly in N. Moreover, if $N(t) \in \mathbb{N}$ for $t \in [0, T_c)$ is so that $N(t) \to \infty$ sufficiently fast as $t \to T_c$, we have

$$\|(1 - \Delta_{x_1})^{1/4}\,\psi_{N(t),t}\|^2 = Tr\,(1 - \Delta)^{1/2}\gamma_{N(t),t}^{(1)} \to \infty \qquad \text{as} \quad t \to T_c^-. \tag{15}$$

To show these two theorems we use the techniques developed in [18]. These techniques were first introduced, in a slightly different context, by Hepp in [10]. They are based on the representation of the many-body system on the Fock space, and on the use of coherent states as initial data. With respect to [18], the main novelty is that, to prove Theorem 3, we need to show convergence not only in the trace norm (as in (14)), but also in the energy norm. In other words, we have to show that, as long as φ_t does not blow up,

$$Tr\left|(1 - \Delta)^{1/4}\left(\gamma_{N,t}^{(1)} - |\varphi_t\rangle\langle\varphi_t|\right)(1 - \Delta)^{1/4}\right| \to 0 \tag{16}$$

as $N \to \infty$. In fact, as pointed out to us by R. Seiringer, the existence of $0 < T_0 \leq T_c$ with the property $\|(1 - \Delta_{x_1})^{1/4}\,\psi_{N(t),t}\|^2 \to \infty$ as $t \to T_0^-$ (with $N(t) \to \infty$ as $t \to T_0^-$) follows from (14) and from the semicontinuity of the kinetic energy (the kinetic energy of the limit is always smaller than the limit of the kinetic energy). However, it is only (16) that allows us to conclude that $T_0 = T_c$ and therefore that the collapse of the many body wave function can really be described by the blow up of the solution of the Hartree equation (9).

3 Dynamics of Bose-Einstein condensates

Since the work of groups around Cornell and Wieman at the University of Colorado, and around Ketterle at MIT, see [1, 2], Bose-Einstein condensation has become accessible to experiments. In these experiments dilute Bose gases are initially trapped by strong magnetic fields. Then, after cooling the gas to very low temperatures (of the order of nano-kelvin), the traps are switched off and the evolution of the gas is observed. To understand these experiments it is important to find an accurate description of the macroscopic properties of the evolution of the condensate.

The trapped Bose gas is described by the Hamiltonian

$$H_N^{\text{trap}} = \sum_{j=1}^{N}(-\Delta_{x_j} + V_{\text{ext}}(x_j)) + \sum_{i<j}^{N} N^2 V(N(x_i - x_j)) \tag{17}$$

acting on the N boson Hilbert space $\mathcal{H}_N = L_s^2(\mathbb{R}^{3N})$. Here V_{ext} is an external potential modeling the magnetic traps, $V \geq 0$ is repulsive and of short range, and the interaction potential $V_N(x) = N^2 V(Nx)$ scales with the number of particles N so that its scattering length is of the order $1/N$. Recall that the scattering length of V is defined as

$$8\pi a_0 = \int dx\, V(x) f(x) \tag{18}$$

where f is the solution of the zero-energy scattering equation

$$\left(-\Delta + \frac{1}{2}V\right) f = 0 \tag{19}$$

with the boundary condition $f(x) \to 1$ as $|x| \to \infty$ (it is then simple to check that $f(x) \simeq 1 - (a_0/|x|) + O(|x|^{-2})$ as $|x| \to \infty$). If f is a solution of (19), it follows by scaling that $f_N(x) = f(Nx)$ solves

$$\left(-\Delta + \frac{1}{2}V_N\right) f_N = 0\,. \tag{20}$$

This implies, in particular, that the scattering length of V_N equals a_0/N.

In [16], Lieb, Seiringer and Yngvason proved that, if E_N denotes the ground state energy of H_N, $E_N/N \to \min_{\varphi:\|\varphi\|=1} \mathcal{E}_{GP}(\varphi)$ as $N \to \infty$. Here

$$\mathcal{E}_{GP}(\varphi) = \int dx\, \left(|\nabla\varphi(x)|^2 + V_{\text{ext}}(x)|\varphi(x)|^2 + 4\pi a_0 |\varphi(x)|^4\right)$$

for all $\varphi \in L^2(\mathbb{R}^3)$ is the so called Gross-Pitaevskii energy functional. In [15], Lieb and Seiringer showed that the ground state of H_N^{trap} exhibits complete condensation into the minimizer ϕ_{GP} of $\mathcal{E}_{GP}$, in the sense that

$$\gamma_N^{(1)} \to |\phi_{GP}\rangle\langle\phi_{GP}| \qquad \text{as} \quad N \to \infty\,, \tag{21}$$

where $\gamma_N^{(1)}$ is the one-particle density of the ground state of H_N^{trap}.

What happens now when the traps are turned off? The system starts to evolve with respect to the translation invariant Hamitonian

$$H_N = \sum_{j=1}^{N} -\Delta_{x_j} + \sum_{i<j}^{N} N^2 V(N(x_i - x_j))\,. \tag{22}$$

It turns out that the macroscopic properties of the resulting evolution can also be approximated by the same Gross-Pitaevskii which successfully describes the properties of the ground state. The following theorem is proven in [4, 5, 7], a series of joint works with L. Erdős and H.-T. Yau.

Theorem 4. *Suppose that $0 \leq V(x) \leq C(1 + x^2)^{-\sigma/2}$ for some $\sigma > 5$. Let $\psi_N \in L^2(\mathbb{R}^{3N})$ be a sequence of N-particle wave functions with $\|\psi_N\| = 1$, so that, as $N \to \infty$, $\gamma_N^{(1)} \to |\varphi\rangle\langle\varphi|$ for an arbitrary $\varphi \in H^1(\mathbb{R}^3)$ (ψ_N exhibits complete condensation) and so that $\langle\psi_N, H_N\psi_N\rangle \leq CN$ (ψ_N has finite energy per particle). Then, for every fixed $t \in \mathbb{R}$, the evolved wave function $\psi_{N,t} = e^{-iH_N t}\psi_N$ still exhibits complete Bose-Einstein condensation, in the sense that*

$$\gamma_{N,t}^{(1)} \to |\varphi_t\rangle\langle\varphi_t| \tag{23}$$

as $N \to \infty$. Here φ_t is the solution of the Gross-Pitaevskii equation

$$i\partial_t \varphi_t = -\Delta\varphi_t + 8\pi a_0 |\varphi_t|^2 \varphi_t \tag{24}$$

with the initial data $\varphi_{t=0} = \varphi$.

Observe that (23) immediately implies that $\gamma_{N,t}^{(k)} \to |\varphi_t\rangle\langle\varphi_t|^{\otimes k}$ for all $k \geq 1$ (this is a general properties of densities converging to rank one projections). This theorem shows that the Gross-Pitaevskii equation can be used to describe the evolution of the condensates in the experiments we discussed above. In the rest of this section, I will explain the main ideas of the proof of Theorem 4. We follow the general strategy introduced in [19] to analyze the mean field limit; we need however to adapt the techniques to the present situation. The main difficulties are a consequence of the fact that (22) does not describe a mean field regime, which is characterized by many weak collisions among the particles. On the contrary, the evolution generated by (22) is characterized by very rare and very strong collisions (particles interact only when they are extremely close, at distances of order $1/N$).

The starting point of our analysis is the study of the time evolution of the reduced density matrices associated with the solution of the Schrödinger equation $\psi_{N,t}$. It turns out that the family $\gamma_{N,t}^{(k)}$, for $k = 1, \ldots, N$, satisfies a hierarchy of N coupled equations, commonly known as the BBGKY hierarchy:

$$i\partial_t \gamma_{N,t}^{(k)} = \sum_{j=1}^{k} \left[-\Delta_{x_j}, \gamma_{N,t}^{(k)} \right] + \sum_{i<j}^{k} \left[N^2 V(N(x_i - x_j)), \gamma_{N,t}^{(k)} \right]$$

$$+ (N-k) \sum_{j=1}^{k} \mathrm{Tr}_{k+1} \left[N^2 V(N(x_j - x_{k+1})), \gamma_{N,t}^{(k+1)} \right] \tag{25}$$

where Tr_{k+1} denotes the partial trace over the degrees of freedom of the $(k+1)$-th particle. From this hierarchy, we try to find equations for the limit points of the reduced densities, as $N \to \infty$. Suppose, for example, that $\gamma_{\infty,t}^{(1)}$ and $\gamma_{\infty,t}^{(2)}$ are limit points of $\gamma_{N,t}^{(1)}$ and, respectively, of $\gamma_{N,t}^{(2)}$. Then, from (25), with $k=1$, we may expect $\gamma_{\infty,t}^{(1)}$ and $\gamma_{\infty,t}^{(2)}$ to satisfy the equation

$$i\partial_t \gamma_{\infty,t}^{(1)} = [-\Delta, \gamma_{\infty,t}^{(1)}] + \mathrm{Tr}_2 \left[b_0 \delta(x_1 - x_2), \gamma_{\infty,t}^{(2)} \right] \tag{26}$$

where we used that, as $N \to \infty$,

$$(N-1)N^2 V(N(x_1 - x_2)) \to b_0 \delta(x_1 - x_2), \quad \text{with} \quad b_0 = \int V(x) dx.$$

It turns out that (26) is not correct. Taking the limit $N \to \infty$ in (25), we ignored the correlations developed by the two point reduced density $\gamma_{N,t}^{(2)}$ on the length scale $1/N$. Using the solution $f_N(x) = f(Nx)$ of the zero energy scattering equation (20) to describe the correlations, we can try to approximate $\gamma_{N,t}^{(1)}, \gamma_{N,t}^{(2)}$, for large but finite N, by

$$\gamma_{N,t}^{(1)}(x_1; x_1') \simeq \gamma_{\infty,t}^{(1)}(x_1; x_1'),$$

$$\gamma_{N,t}^{(2)}(x_1, x_2; x_1', x_2') \simeq f_N(x_1 - x_2) f_N(x_1' - x_2') \gamma_{\infty,t}^{(2)}(x_1, x_2; x_1', x_2'). \tag{27}$$

Inserting this ansatz in (25) (for $k=1$), and using the definition (18) of the scattering length, we easily find another (and, this time, correct) equation for the limit points $\gamma_{\infty,t}^{(1)}, \gamma_{\infty,t}^{(2)}$:

$$i\partial_t \gamma_{\infty,t}^{(1)} = \left[-\Delta, \gamma_{\infty,t}^{(1)} \right] + 8\pi a_0 \mathrm{Tr}_2 \left[\delta(x_1 - x_2), \gamma_{\infty,t}^{(2)} \right]. \tag{28}$$

Note that, $f_N \to 1$ as $N \to \infty$ (in a weak sense). For this reason, f_N only plays an important role when it is coupled with the (very) singular potential V_N. Similarly to (28), starting from (25) for $k>1$ and considering the limit $N \to \infty$ (taking into account the short scale correlation structure developed by $\gamma_{N,t}^{(k+1)}$), we conclude that an arbitrary limit point $\{\gamma_{\infty,t}^{(k)}\}_{k \geq 1}$ of the sequence of reduced density matrices $\{\gamma_{N,t}^{(k)}\}_{k=1}^{N}$ satisfies the infinite hierarchy of equations

$$i\partial_t \gamma_{\infty,t}^{(k)} = \sum_{j=1}^{k} \left[-\Delta_{x_j}, \gamma_{\infty,t}^{(k)} \right] + 8\pi a_0 \sum_{j=1}^{k} \mathrm{Tr}_{k+1} \left[\delta(x_j - x_{k+1}), \gamma_{\infty,t}^{(k+1)} \right]. \tag{29}$$

It is then useful to observe that the factorized densities

$$\gamma_{\infty,t}^{(k)} = |\varphi_t\rangle\langle\varphi_t|^{\otimes k} \tag{30}$$

solve (29) if and only if φ_t is a solution of the Gross-Pitaevskii equation (24).

The strategy to show Theorem 4 consists therefore of the following steps.

1) Show the compactness of the sequence of families $\{\gamma_{N,t}^{(k)}\}_{k=1}^{N}$ with respect to an appropriate weak topology.
2) Prove that the reduced densities $\gamma_{N,t}^{(k+1)}$, $k \geq 1$, have a short scale correlation structure which can be described, in good approximation, by the solution f_N of the zero energy scattering equation (20).
3) Using the short scale structure developed by $\gamma_{N,t}^{(k+1)}$, show that any limit point of the sequence $\{\gamma_{N,t}^{(k)}\}_{k=1}^{N}$ is a solution of the infinite hierarchy (29).
4) Prove the uniqueness of the solution of the infinite hierarchy (29).

Steps 1-4 conclude the proof of Theorem 4 because a compact sequence with at most one limit point must be convergent and therefore (30) must be its limit. Details can be found in [4, 5, 6, 7]. Recently, an alternative proof of Theorem 4 was proposed by Pickl in [17]. Pickl's approach also allows for the presence of (possibly time-dependent) external potentials in the Hamiltonian (22) (but requires stronger conditions on the initial wave function).

References

1. M.H. Anderson, J.R. Ensher, M.R. Matthews et al: Science **269**, 198 (1995).
2. K. B. Davis, M. -O. Mewes, M. R. Andrews et al: Phys. Rev. Lett. **75**, 3969 (1995).
3. A. Elgart, B. Schlein: Commun. Pure Appl. Math. **60**, 500 (2007).
4. L. Erdős, B. Schlein, H.-T. Yau: Invent. Math. **167**, 515 (2007).
5. L. Erdős, B. Schlein, H.-T. Yau: Ann. Math. **172**, 291 (2010).
6. L. Erdős, B. Schlein, H.-T. Yau: Phys. Rev. Lett. **98** (2007), 040404.
7. L. Erdős, B. Schlein, H.-T. Yau: J. Amer. Math. Soc. **22**, 1099 (2009).
8. L. Erdős, H.-T. Yau: Adv. Theor. Math. Phys. **5**, 1169 (2001).
9. J. Fröhich, E. Lenzmann: Comm. Pure Appl. Math. **60**, 1691 (2007).
10. K. Hepp: Comm. Math. Phys. **35**, 265 (1974).
11. A. Knowles, P. Pickl: Comm. Math. Phys. **298**, 101 (2010).
12. E. Lenzmann: Math. Phys. Anal. Geom. **10**, 43 (2007).
13. E. H. Lieb, R. Seiringer: Phys. Rev. Lett. **88** (2002), 170409-1-4.
14. E. H. Lieb, R. Seiringer, J. Yngvason: Phys. Rev. A **61** (2000), 043602.
15. E. H. Lieb, H.-T. Yau: Comm. Math. Phys. **112**, 147 (1987).
16. A. Michelangeli, B. Schlein: Dynamical Collapse of Boson Stars. Preprint arXiv:1005.3135.
17. P. Pickl: GP derivation. A simple derivation of mean field limits for quantum systems. Preprint arXiv:0907.4464.
18. I. Rodnianski, B. Schlein: Comm. Math. Phys. **291**, 31 (2009).
19. H. Spohn: Rev. Mod. Phys. **52**, 569 (1980), no. 3.

Bose gases in rapid rotation

Jakob Yngvason

Faculty of Physics, University of Vienna, Boltzmanngasse 5, A 1090 Vienna, and
ESI, Boltzmanngasse 9, A 1090 Vienna
jakob.yngvason@univie.ac.at

Abstract. Some recent rigorous results on the effect of rapid rotation on ultracold Bose gases are reviewed. These include a phase transition to a giant vortex state in anharmonic traps and a derivation of a mean field description of a gas in a harmonic trap at rotation frequencies close to the trap frequency.

1 Quantum vortices

Consider a fluid with velocity field $\mathbf{v}(\mathbf{x})$. The circulation around a closed loop $\mathcal{C}$ enclosing a domain $\mathcal{D}$ is, by Stokes' Theorem,

$$\oint_{\mathcal{C}} \mathbf{v} \cdot d\ell = \int_{\mathcal{D}} (\nabla \times \mathbf{v}) \cdot \mathbf{n} dS. \tag{1}$$

Hence nonzero circulation requires that the *vorticity* $\nabla \times \mathbf{v}$ is nonzero somewhere in $\mathcal{D}$. A region where $\nabla \times \mathbf{v} \neq 0$ is called a *vortex*.

A superfluid is described by a complex valued function ("order parameter") ψ satisfying a nonlinear Schrödinger Equation, the *Gross-Pitaevskii equation* (see Section 3 below). The phase of ψ determines the velocity: If $\psi = e^{i\varphi}|\psi|$ then the continuity equation that follows the GP equation implies

$$\mathbf{v} = (\hbar/m)\nabla\varphi \tag{2}$$

where m is the mass of the quantum particles. Since ψ is single valued we have $\oint_{\mathcal{C}} \nabla\varphi \cdot d\ell = n\,2\pi$ with $n \in \mathbb{Z}$, so

$$\oint_{\mathcal{C}} \mathbf{v} \cdot d\ell = n(h/m). \tag{3}$$

Thus, in a superfluid, *vorticity is quantized* in units of h/m. This basic fact was probably first realized by L. Onsager, see [19], p. 693. The appearance of vortices is accompanied by a vanishing of the wave function and a corresponding singularity of the phase. Where $|\psi(\mathbf{r})| \neq 0$ and the phase is nonsingular we have $\nabla \times \mathbf{v} = (\hbar/m)\nabla \times \nabla\varphi = 0$.

The physics of quantized vortices in dilute, trapped ultracold gases has been intensely studied, both experimentally and theoretically, over the past decade. Excellent reviews, containing many further references, can be found in [1, 9, 4]. The present notes are devoted to some recent results on the mathematical physics of Bose gases in rapid rotation, in particular from the papers [8, 7, 17]. The general mathematical background is described in [18].

2 The basic many-body hamiltonian

The quantum mechanical Hamiltonian for N spinless bosons with a pair interaction potential v and external potential V in a rotating frame with angular velocity $\boldsymbol{\Omega}$ is

$$H = \sum_{j=1}^{N} \left(-\tfrac{1}{2}\nabla_j^2 + V(\mathbf{x}_j) - \mathbf{L}_j \cdot \boldsymbol{\Omega} \right) + \sum_{1 \leq i < j \leq N} v(|\mathbf{x}_i - \mathbf{x}_j|). \tag{4}$$

Here $\mathbf{x}_j \in \mathbb{R}^3$, $j = 1, \dots, N$ is the position and $\mathbf{L}_j = -i\,\mathbf{x}_j \times \nabla_j$ the angular momentum operator for particle j. Units have been chosen so that $\hbar = m = 1$ and thus $h/m = 2\pi$. The pair interaction potential v is assumed to be radially symmetric, of short range and repulsive.

The Hamiltonian H operates on *symmetric* functions in $L^2(\mathbb{R}^{3N})$. It can alternatively be written in the form

$$H = \sum_{j=1}^{N} \left(\tfrac{1}{2}[i\nabla_j + \mathbf{A}(\mathbf{x}_j)]^2 + V(\mathbf{x}_j) - \tfrac{1}{2}\Omega^2 r_j^2 \right) + \sum_{1 \leq i < j \leq N} v(|\mathbf{x}_i - \mathbf{x}_j|) \tag{5}$$

with

$$\mathbf{A}(\mathbf{x}) = \boldsymbol{\Omega} \times \mathbf{x} = \Omega r\, \mathbf{e}_\theta \tag{6}$$

where $\Omega = |\boldsymbol{\Omega}|$, r is the distance from the rotation axis, and $\mathbf{e}_\theta$ the unit vector in the angular direction. This way of writing the Hamiltonian corresponds to the splitting of the rotational effects into Coriolis and centrifugal forces. The Coriolis forces have formally the same effect as a magnetic field $\mathbf{B} = \nabla \times \mathbf{A} = 2\boldsymbol{\Omega}$.

When discussing systems in rapid rotation it is necessary to distinguish between *harmonic* and *anharmonic* traps. If V is harmonic in the directions perpendicular to $\boldsymbol{\Omega}$, i.e.,

$$V(\mathbf{x}) = \tfrac{1}{2}\Omega_{\text{trap}} r^2 + V^{\|}(z) \tag{7}$$

with $V^{\|}$ a confining potential in the direction of $\boldsymbol{\Omega}$, then stability requires $\Omega < \Omega_{\text{trap}}$. *Rapid rotation* means here that $\Omega \to \Omega_{\text{trap}}$ from below. If V is anharmonic i.e., increases *faster than quadratically* in the directions perpendicular to $\boldsymbol{\Omega}$, e.g. $V(\mathbf{x}) \sim r^s$ with $s > 2$, then rapid rotation means simply $\Omega \to \infty$.

3 Gross-Pitaevskii equation

In *dilute* gases the effect of the interaction potential v is, to a very good approximation, subsumed in a single parameter, the *scattering length* a, and the many-body system can be described by a wave function on $\mathbb{R}^3$ rather than on $\mathbb{R}^{3N}$. This was rigorously established in [16] for the non-rotating case and in [15] for rotating gases as well. The basic fact proved in [15] can be stated as follows: If $N \to \infty$ with Na and $\mathbf{\Omega}$ fixed then *Bose-Einstein condensation* takes place in the the ground state of (4) and the wave function of the condensate satisfies the *Gross-Pitaevskii* (GP) *equation*

$$\left\{ \tfrac{1}{2}(\mathrm{i}\nabla + \mathbf{A})^2 + (V - \tfrac{1}{2}\Omega^2 r^2) + 2\pi Na|\psi|^2 \right\} \psi = \mu\psi . \tag{8}$$

Here μ is a chemical potential determined by the normalization $\int_{\mathbb{R}^3} |\psi|^2 = 1$. The GP equation is obtained by minimizing the *GP energy functional*

$$\mathcal{E}^{\mathrm{GP}}[\psi] = \int_{\mathbb{R}^3} \left\{ \tfrac{1}{2}|\nabla\psi|^2 + V|\psi|^2 - \psi^* \, \mathbf{\Omega} \cdot \mathbf{L}\psi + \pi Na|\psi|^4 \right\} d\mathbf{x}$$

$$= \int_{\mathbb{R}^3} \left\{ \tfrac{1}{2}|(\mathrm{i}\nabla + \mathbf{A})\psi|^2 + (V - \tfrac{1}{2}\Omega^2 r^2)|\psi|^2 + \pi Na|\psi|^4 \right\} d\mathbf{x}$$

under the stated normalization condition for ψ.

The GP minimization problem has *two parameters*, Ω and Na. In anharmonic traps we shall consider it in the *asymptotic regime* where *both* these parameters are large. It is customary and convenient to introduce

$$\varepsilon := (\pi Na)^{-1/2} \tag{9}$$

which is small if Na is large.

In harmonic traps with $\Omega \to \Omega_{\mathrm{trap}}$ it turns out to be appropriate to restrict the wave functions to the *Lowest Landau Level* of the magnetic Hamiltonian $\tfrac{1}{2}(\mathrm{i}\nabla + \mathbf{A})^2$. We shall have something to say about this regime in Sections 6 and 7 both for large and small values of Na.

The rigorous derivation of the GP equation from the many-body problem in the rotating case in [15] is carried out for Ω and ε fixed. For rapid rotation the GP description may break down both in harmonic and anharmonic traps if these parameters are not kept fixed as $N \to \infty$. The exact limitations can be conjectured but are still not completely established. A step in this direction for anharmonic traps was proved in [3]:

If $N \to \infty$ and $\Omega \to \infty$ but the gas remains dilute (in the sense that the mean density is $\ll a^{-3}$) the *TF approximation*, i.e., the GP energy without the kinetic term $\tfrac{1}{2}|(\mathrm{i}\nabla + \mathbf{A})\psi|^2$, gives the leading term in the ground state many-body energy per particle as a function of Ω and ε.

4 GP theory for rapid rotation, anharmonic traps

Consider, for simplicity, a 2D 'flat', circular trap with radius 1. The GP energy functional is then

$$\mathcal{E}^{\mathrm{GP}}[\psi] = \int_{\mathcal{B}} \left\{ \tfrac{1}{2} |(\mathrm{i}\nabla + \mathbf{A})\psi|^2 - \tfrac{1}{2}\Omega^2 r^2 |\psi|^2 + \frac{1}{\varepsilon^2}|\psi|^4 \right\} d\mathbf{r} \qquad (10)$$

with $\mathcal{B}$ the unit disc and $\mathbf{A}(\mathbf{r}) = \Omega\, r\, \mathbf{e}_\theta$, $\mathbf{r} \in \mathbb{R}^2$.

It can be proved that if $\Omega \le \Omega_1 |\log\varepsilon| + O(\log|\log\varepsilon|)$ for a certain constant Ω_1 any minimizer has is a *finite* number of vortices, even as $\varepsilon \to 0$ [12, 13, 20, 2]. For larger Ω the number of vortices is *unbounded* as $\varepsilon \to 0$.

If $\Omega = O(1/\varepsilon)$ the centrifugal term $-(\Omega^2/2)r^2|\psi(\mathbf{r})|^2$ and the interaction term $(1/\varepsilon^2)|\psi(\mathbf{r})|^4$ are comparable in size. The kinetic energy term $\tfrac{1}{2}|(\mathrm{i}\nabla + \mathbf{A}(\mathbf{r}))\psi(\mathbf{r})|^2$ is formally also of order $1/\varepsilon^2$ if $\Omega \sim 1/\varepsilon$. However, it turns out that its contribution to the energy is, in fact, of *lower order*, namely $\sim \Omega|\log\varepsilon|$, because *a lattice of vortices emerges* as $\varepsilon \to 0$. The velocity field generated by the vortices compensates partly that of $\mathbf{A}(\mathbf{r}) = \Omega r\, \mathbf{e}_\theta$. The energy asymptotics to subleading order is given by the following theorem proved in [8]:

Theorem 1 (Energy asymptotics in a flat trap) *Let E^{GP} denote the GP energy, i.e., the minimum of the GP energy functional (10). Let E^{TF} denote the minimal energy of the GP functional without the kinetic term. If $|\log\varepsilon| \ll \Omega \ll 1/\varepsilon$, then*

$$E^{\mathrm{GP}} = E^{\mathrm{TF}} + \tfrac{1}{2}\Omega|\log(\varepsilon^2\Omega)|(1 + o(1)). \qquad (11)$$

If $1/\varepsilon \lesssim \Omega \ll 1/(\varepsilon^2|\log\varepsilon|)$ then

$$E^{\mathrm{GP}} = E^{\mathrm{TF}} + \tfrac{1}{2}\Omega|\log\varepsilon|(1 + o(1)). \qquad (12)$$

The limiting velocity $1/(\varepsilon^2|\log\varepsilon|)$ for the validity of (12) may appear strange at this point but it has a natural explanation as discussed in Section 5 below.

4.1 An electrostatic analogy

The logarithmic terms in (11) and (12) are the contributions from a lattice of vortices and can be understood by the following heuristic considerations. Let us write points $\mathbf{r} = (x, y) \in \mathbb{R}^2$ as complex numbers, $\zeta = x + \mathrm{i}y$, and consider a lattice of points ζ_i. Placing a vortex of degree 1 at each point ζ_i leads to a trial function for the GP energy of the form

$$\psi(\mathbf{r}) = f(\mathbf{r}) \exp\{\mathrm{i}\varphi(\mathbf{r})\} \qquad (13)$$

where f is real valued and

$$\exp\{i\varphi(\mathbf{r})\} = \prod_i \frac{\zeta - \zeta_i}{|\zeta - \zeta_i|}. \tag{14}$$

Now

$$|(i\nabla + \mathbf{A})\psi|^2 = |\nabla f|^2 + f^2|\mathbf{A} - \nabla\varphi|^2 \tag{15}$$

and by the Cauchy-Riemann equations for the complex logarithm we can write

$$|\mathbf{A} - \nabla\varphi|^2 = |\Omega\, r\mathbf{e}_r - \nabla\chi|^2 \tag{16}$$

where

$$\chi(\mathbf{r}) = \sum_i \log|\mathbf{r} - \mathbf{r}_i|. \tag{17}$$

But

$$\mathbf{E}(\mathbf{r}) \equiv \Omega\, r\mathbf{e}_r - \nabla\chi(\mathbf{r}) \tag{18}$$

has a simple physical interpretation: It can be regarded as an 'electric field' generated by a *uniform charge distribution* of density Ω/π together with *unit 'charges' of opposite sign* at the positions of the vortices, $\mathbf{r}_i$.

We now distribute the vortices over the unit disk so that the vorticity per unit area is Ω/π. (This is really $2\Omega \cdot m/h$.) Thus every vortex $\mathbf{r}_i$ sits at the center of lattice cell Q_i of area $|Q_i| = \pi/\Omega$, surrounded by a uniform charge distribution of the opposite sign so that the total charge in the cell is zero. *If* the cells were disc-shaped, then *Newton's theorem* would imply that the 'electric field' generated by the cell would vanish outside the cell, i.e, there would be *no interaction between the cells*.

The cells can, of course, not be disc shaped, but the closest approximation to that are *hexagonal cells*, giving the optimal energy. The vortices then sit on a *triangular lattice*, and this is, indeed, what is observed experimentally. The interaction between the cells, although not zero, is small because the cells have no dipole moment.

The energy of a single vortex is to leading order $\rho(\mathbf{r}_i)|\log(t^2\Omega)|$ where ρ is the bulk density (i.e., the density of the gross profile without vortices) at $\mathbf{r}_i$, and t is the radius of a small disk (vortex ball) around $\mathbf{r}_i$ where ψ is essentially zero. Optimizing the sum of the kinetic energy of a vortex and the increase in interaction energy due to the redistribution of the density accompanied with the creation of the vortex gives $t \sim \varepsilon/\rho^{1/2}$. If $\Omega \lesssim 1/\varepsilon$ we have $\rho = O(1)$ while for $\Omega \gg 1/\varepsilon$ the centrifugal forces dominate the repulsive interaction forces and the density becomes *concentrated in a thin annulus* of size $\sim (\varepsilon\Omega)^{-1}$. Here $\rho \sim (\varepsilon\Omega)$. Inserting the corresponding values for t and summing up the vortex contributions gives the right hand sides of (11) and (12) respectively as upper bounds. The lower bounds are obtained with the help of techniques from Ginzburg-Landau theory for which [23] is an excellent reference.

5 Emergence of a 'giant vortex'

In the regime $\Omega \gg 1/\varepsilon$ the thin annulus $\mathcal{A}$ of size $\sim (\varepsilon\Omega)^{-1}$ where the bulk of the condensate sits may still contain a lattice of vortices, and also the total vorticity of the 'hole' $\mathcal{B} \setminus \mathcal{A}$, where the density is vanishingly small, is equal to the area of the 'hole' times the vortex density Ω/π. For

$$\Omega > \frac{\Omega_c}{\varepsilon^2 |\log \varepsilon|} \tag{19}$$

with a certain $\Omega_c > 0$, however, a *phase transition* occurs: Vortices disappear from the annulus and *all* the vorticity is concentrated in the 'hole'. This is called a *Giant Vortex* [10, 11].

A heuristic explanation for the transition at $\Omega \sim 1/(\varepsilon^2 |\log \varepsilon|)$ can be given by exploiting the electrostatic analogy: A variational *ansatz* for ψ of the form

$$\psi(\mathbf{r}) = f(\mathbf{r}) \exp(\mathrm{i}\hat{\Omega}\theta) \tag{20}$$

with a real valued function f is optimal if

$$\hat{\Omega} = \Omega - O(\varepsilon^{-1}). \tag{21}$$

The 'electric field' generated by a charge $\hat{\Omega}$ at the origin exactly cancels, in the annulus of thickness $(\varepsilon\Omega)^{-1}$, the 'electric field' generated in the annulus by the uniform charge density Ω/π of the 'hole' (by Newton's theorem). However, there is still a residual field in the annulus generated by the 'charge' in the annulus itself which is

$$\text{charge density} \times \text{area of annulus} \sim \Omega \times (\varepsilon\Omega)^{-1} = \varepsilon^{-1}. \tag{22}$$

The electrostatic energy of this residual charge distribution is $\sim \varepsilon^{-2}$. Creating a vortex in the annulus neutralizes one charge unit and thus reduces the electrostatic energy by ε^{-1}. In other words,

$$\text{gain by creating a single vortex} \sim \frac{1}{\varepsilon}. \tag{23}$$

On the other hand, the *cost* of a vortex is $\sim \rho |\log \varepsilon|$, and we have $\rho \sim (\varepsilon\Omega)$, so

$$\text{cost of a single vortex} \sim \varepsilon\Omega |\log \varepsilon|. \tag{24}$$

Gain and cost are thus comparable if

$$\Omega \sim \frac{1}{\varepsilon^2 |\log \varepsilon|}. \tag{25}$$

If Ω is smaller it still pays to have vortices also in the annulus, but if Ω is larger, the cost outweighs the gain and there are *no vortices in the annulus*. In other words: If $\Omega > (\text{const.})\,\varepsilon^{-2}|\log \varepsilon|^{-1}$ *all vorticity originates in the region where the density is vanishingly small*.

A mathematical proof substantiating these heuristic considerations is, however, far from simple. It is given in [7] and can be stated as follows:

Theorem 2 (Giant vortex theorem) *Suppose* $\Omega = \Omega_0(\varepsilon^2|\log\varepsilon|)^{-1}$. *If* $\Omega_0 > (3\pi)^{-1}$, *then* ψ^{GP} *has, for small* ε, *no zeros in the annulus where the bulk of the density is concentrated. Moreover, the energy* E^{GP} *is precisely approximated by the giant vortex ansatz* (20), *up to terms* $O(\varepsilon^{-1/2}|\log\varepsilon|^{3/2})$ *that are much smaller than the energy of a single vortex in the annulus which is* $O(\varepsilon^{-1})$.

Further rigorous results on giant vortices, also in more general anharmonic traps than the flat trap considered here, are proved in [21, 22, 5, 6].

6 Rapid rotation, harmonic traps

In a harmonic trap potential $V(\mathbf{r}) = \frac{1}{2}\Omega_{\mathrm{trap}}r^2 + V^{\|}(z)$ the rotational velocity Ω can not exceed Ω_{trap}. As Ω approaches the critical velocity Ω_{trap} from below interesting phenomena appear:

- The 3D system becomes *effectively 2D* and its ground state confined to the *Lowest Landau Level* (LLL) of the magnetic Hamiltonian $\frac{1}{2}(i\nabla + \mathbf{A})^2$. If the interaction is not too strong a lattice of vortices appears as $\Omega \to \Omega_{\mathrm{trap}}$.
- If Ω is sufficiently close to Ω_{trap} and the interaction not too weak, *the vortex lattice 'melts'* and the many-body state becomes *strongly correlated*, with features analogous to those encountered in the *Fractional Quantum Hall Effect* for charged fermions.

For the mathematical description of LLL physics the appropriate one-particle Hilbert space is the *Bargmann space* $\mathcal{H}$ of *analytic* functions φ such that

$$\langle\varphi,\varphi\rangle = \int |\varphi(z)|^2 \exp(-|z|^2)\,\mathrm{d}^2z < \infty \tag{26}$$

where d^2z denotes the Lebesgue measure on $\mathbb{C}$ (regarded as $\mathbb{R}^2$). This is a Hilbert space with scalar product

$$\langle\varphi,\psi\rangle = \int \overline{\varphi(z)}\psi(z) \exp(-|z|^2)\,\mathrm{d}^2z. \tag{27}$$

On $\mathcal{H}$ the angular momentum operator is $L = z\partial$.

For the the bosonic N-body problem the corresponding Hilbert space is

$$\mathcal{H}_N = \mathcal{H}^{\otimes_{\mathrm{symm}}^N}, \tag{28}$$

i.e., the space of symmetric, analytic functions ϕ of $z_1, \ldots, z_N$ such that

$$\int_{\mathbb{C}^N} |\phi(z_1, \ldots, z_N)|^2 \exp(-|z_1|^2 - \cdots - |z_N|^2)\,\mathrm{d}^2z_1 \cdots \mathrm{d}^2z_N < \infty. \tag{29}$$

The *interaction* can be taken into account by a suitable projection of the potential energy operator $\sum_{1 \leq i < j \leq N} v(|\mathbf{x}_i - \mathbf{x}_j|)$ onto $\mathcal{H}_N$. In fact, for short range interaction potentials it was proved in [14] that one can replace $v(\mathbf{x}_i - \mathbf{x}_j)$ by $g\delta(z_i - z_j)$ with a coupling constant $g \sim a > 0$. A delta-function potential does not in general make sense in a higher dimension than 1, but for analytic functions it is perfectly acceptable:

It is sufficient to consider $N = 2$. Define an operator δ_{12} on $\mathcal{H}_2$ by

$$\delta_{12}\phi(z_1, z_2) = \frac{1}{2\pi}\phi\big(\tfrac{1}{2}(z_1 + z_2), \tfrac{1}{2}(z_1 + z_2)\big). \tag{30}$$

Using the analyticity of ϕ it is easy to see that

$$\int_{\mathbb{C}^2} \overline{\phi(z_1, z_2)}(\delta_{12}\phi)(z_1, z_2) \exp(-|z_1|^2 - |z_2|^2)\,\mathrm{d}^2 z_1\,\mathrm{d}^2 z_2 =$$

$$\int_{\mathbb{C}} |\phi(z, z)|^2 \exp(-2|z|^2)\,\mathrm{d}^2 z, \quad (31)$$

i.e., δ_{12} has formally the same effect as $\delta(z_1 - z_2)$. Note that δ_{12} is a bounded operator; in fact $(2\pi)\delta_{12}$ is a projector.

The many-body Hamiltonian on $\mathcal{H}_N$ can now be written [14]

$$H_{N,\omega,g} = \omega \sum_{i=1}^{N} L_i + g \sum_{1 \leq i < j \leq N} \delta_{ij} \equiv \omega \hat{L} + g\hat{I} \tag{32}$$

with $\omega := \Omega_{\mathrm{trap}} - \Omega$. The task is to describe its ground state(s) for large N. We denote its ground state energy by $E(N, \omega, g)$.

The ground state energy is determined by the lower boundary of the joint spectrum of the commuting operators $\hat{L}$ and $\hat{I}$, i.e, by the *Yrast curve*

$$I(L) = \text{lowest eigenvalue of } \hat{I} \text{ for a given eigenvalue } L \text{ of } \hat{L}. \tag{33}$$

The ground state energy is then given by the Legendre transform

$$E(N, \omega, g) = \inf_L\{\omega L + gI(L)\}, \tag{34}$$

It is expected, but not proved yet, that as $N \to \infty$ the Yrast curve becomes convex and monotonously decreasing with $I(L) = O(N)$ if $L = O(N^2)$. It is known that $I = 0$ for $L = N(N-1)$; the corresponding wave function is the *Laughlin wave function*

$$\phi(z_1, \ldots, z_N) = \prod_{i<j}(z_i - z_j)^2 \exp(-\textstyle\sum_k |z_k|^2). \tag{35}$$

This function was first introduced in connection with the *Fractional Quantum Hall Effect* for electrons (with exponent 3 rather than 2). We define the *filling factor* of a state of angular momentum L (for large N) as

$$\nu = \frac{N^2}{2L}. \tag{36}$$

The Laughlin state has filling factor $\frac{1}{2}$. For all lower filling factors the interaction energy is also zero.

7 The GP approximation in the LLL

For filling factors between $\frac{1}{2}$ and ∞ *upper bounds* on $E(N, \omega, g)$ based on a variety of trial wave functions are known [4]. But matching *lower bounds* are largely missing. For $\nu \to \infty$ (i.e., $L \ll N^2$), however, a complete solution can be given.

Define the *GP energy functional in the LLL* for functions $\varphi \in \mathcal{H}$ by

$$\mathcal{E}^{\mathrm{GP}}[\varphi] = \omega \langle \varphi, L\varphi \rangle + \frac{g}{2} \int_{\mathbb{C}} |\varphi(z)|^4 \exp(-2|z|^2)\, \mathrm{d}^2 z. \tag{37}$$

The GP energy is

$$E^{\mathrm{GP}}(N, \omega, g) = \inf\{\mathcal{E}^{\mathrm{GP}}[\varphi] \ : \ \textstyle\int |\varphi(z)|^2 \exp(-|z|^2)\, \mathrm{d}^2 z = N\}. \tag{38}$$

It satisfies the simple scaling relation

$$E^{\mathrm{GP}}(N, \omega, g) = N\omega E^{\mathrm{GP}}(1, 1, Ng/\omega) \tag{39}$$

and thus has one parameter, the *GP parameter*

$$Ng/\omega. \tag{40}$$

The GP energy is in any case an *upper bound* to $E(N, \omega, g)$. This is seen by using trial functions of the form

$$\phi(z_1, \ldots, z_N) = \varphi(z_1) \cdots \varphi(z_N) \tag{41}$$

for the N-body Hamiltonian. The expectation value is

$$\langle \phi, H\phi \rangle = N\omega \left(\langle \varphi, L\varphi \rangle + \frac{(N-1)g}{2\omega} \int_{\mathbb{C}} |\varphi(z)|^4 \exp(-2|z|^2)\, \mathrm{d}^2 z \right). \tag{42}$$

For the *lower bound* we have [17]:

Theorem 3 (Lower bound for large filling factor) *For every $c > 0$ there is a $C < \infty$ such that*

$$E(N, \omega, g) \geq E^{\mathrm{GP}}(N, \omega, g)(1 - C(g/N\omega)^{1/10}) \tag{43}$$

provided $gN/\omega > c$.

Remark: The effective radius R of the GP minimizer can be estimated by equating ωR^2 with gNR^{-2}, i.e., $R \sim (Ng/\omega)^{1/4}$. The angular momentum per particle is $L_1 \sim R^2 \sim (Ng/\omega)^{1/2}$ and the filling factor is $\nu \sim N^2/(NL_1) \sim (N\omega/g)^{1/2}$. Thus $g/(N\omega) \to 0$ corresponds to $\nu \to \infty$. Note also that the GP parameter gN/ω need not be fixed for the remainder in (43) to be small; it suffices that it is $\ll N^2$.

References

1. A. Aftalion: Progr. Nonlinear Differential Equations Appl. **67**, Birkhäuser, Basel, 2006.
2. A. Aftalion, R.L. Jerrard, J. Royo-Letelier: Non Existence of Vortices in the Small Density Region of a Condensate. Preprint arXiv:1008.4801v1.
3. J.-B. Bru, M. Correggi, P. Pickl, J. Yngvason: Comm. Math. Phys. **280**, 517 (2008).
4. N.R. Cooper: Adv. Phys. **57**, 539 (2008).
5. M. Correggi, F. Pinsker, N. Rougerie, J. Yngvason, Critical Speeds in the Gross-Pitaevskii Theory on a Disc with Dirichlet Boundary Conditions, in preparation.
6. M. Correggi, F. Pinsker, N. Rougerie, J. Yngvason, On the transition to the giant vortex state in a superfluid in a rotating anharmonic trap, in preparation.
7. M. Correggi, N. Rougerie, J. Yngvason: The Transition to a Giant Vortex Phase in a Fast Rotating Bose-Einstein Condensate. Preprint arXiv:1005.0686.
8. M. Correggi, J. Yngvason: J. Phys. A: Math. Theor. **41**, 445002 (2008).
9. A.L. Fetter: Rev. Mod. Phys. **81**, 647 (2009).
10. A.L. Fetter: Phy. Rev. A **64**, 063608 (2001).
11. U.R. Fischer, G. Baym: Phys. Rev. Lett. **90**, 140402 (2003).
12. R. Ignat, V. Millot: J. Funct. Anal. **233**, 260 (2006).
13. R. Ignat, V. Millot: Rev. Math. Phys. **18**, 119 (2006).
14. M. Lewin, R. Seiringer: J. Stat. Phys. **137**, 1040 (2009).
15. E.H. Lieb, R. Seiringer: Comm. Math. Phys. **264**, 505 (2006).
16. E.H. Lieb, R. Seiringer, J. Yngvason: Phys. Rev. A. **61**, 043602 (2000).
17. E.H. Lieb, R. Seiringer, J. Yngvason: Phys. Rev. A **79**, 063626 (2009).
18. E.H. Lieb, R. Seiringer, J.P. Solovej, J. Yngvason: *The Mathematics of the Bose Gas and its Condensation*, Oberwolfach Seminar Series **34**, Birkhäuser, Basel, (2005)
19. L. Onsager:, *Collected Works*, Editors P.C. Hemmer, H. Holden, S. Kjelstrup Ratkje, World Scientific, Singapore, 1996.
20. T. Rindler-Daller: Physica A, **387**, 1851 (2008).
21. N. Rougerie: The Giant Vortex State for a Bose-Einstein Condensate in a Rotating Anharmonic Trap: Extreme Rotation Regimes. Preprint arXiv:0809.1818v4.
22. N. Rougerie: Vortex Rings in Fast Rotating Bose-Einstein Condensates. Preprint arXiv:1009.1982v1.
23. E. Sandier, S. Serfaty: Progr. Nonlinear Differential Equations Appl. **70**, Birkhäuser, Basel, 2007.

Four-mode heat – Second Josephson oscillations in a small Bose-Hubbard system

Martin P. Strzys and James R. Anglin

OPTIMAS Research Center and Fachbereich Physik,
Technische Universität Kaiserslautern, D–67653 Kaiserslautern, Germany
strzys@physik.uni-kl.de

Abstract. A four-mode Bose-Hubbard model with two highly differing tunneling rates is considered as a model for two quantum systems in thermal contact. In addition to coherent particle exchange a novel slow second Josephson mode, which is not predicted by linear Bogoliubov theory, can be identified by a series of Holstein-Primakoff transformations. This energy exchange mode can be interpreted as heat exchange between the subsystems, is in close analogy to second sound in liquid helium, and may shed light on the emergence of thermodynamics in mesoscopic systems.

1 Motivation

The emergence of thermodynamics in mesoscopic systems is still a field of ongoing research [1, 2, 3, 4, 5, 6]. Thermodynamics is predicated on statistical averages and the introduction of ensembles, which has indeed proven to be a very powerful technique, especially dealing with equilibrium properties and phase transitions. On the other hand, the natural way of dealing with smaller systems of less complexity is the Hamiltonian framework. While forces and work find their original place in the Hamiltonian, the other fundamental ingredient of thermodynamics, namely heat, inescapably needs some averaging process to be well established, which in the end leads to the ensemble formalism. Nevertheless this is only the second step. The canonical derivation involves time averages, which in general are technically not feasible, as a first step. Subsequently ensembles are used as proxies, justified by the quasi-ergodic hypothesis [7]. But in the end both perceptions should be equivalent and it should be possible to deduce the thermodynamic properties directly by taking time averages or rather by deriving some effective adiabatic theory. Quantities like heat would then be deduced directly from the Hamiltonian.

The message behind this although is, that heat can be defined by time-scale separation: Energy stored in degrees of freedom with an evolution to fast to observe or control is averaged over by this procedure and thus usually perceived as heat. Furthermore, energy can certainly be exchanged by fast and slow degrees of freedom. This fact just reflects the first law of thermodynamics, heat can be transformed to work and vice versa. Thus, to investigate the onset of thermodynamics in the mesoscopic regime one needs a system with simple but sufficiently complex dynamics. One minimalist model for two quantum systems in thermal contact can be realized by two coupled ideal Josephson junctions [8].

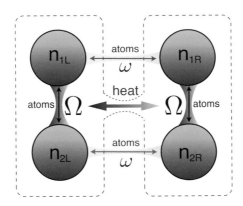

Fig. 1. Bose-Hubbard system as model for thermal contact.

Each subsystem is realized by a two-mode Bose-Hubbard (BH) system. Its characteristic timescale—the Josephson frequency—is constructed to be the shortest one in the entire system and only much longer time scales are under observation and control. Each subsystem then has, in isolation, exactly two conserved quantum numbers: particle number and energy. For a fixed number of particles the energy may still be varied independently by exciting the Josephson mode. If in an adiabatic theory the rapid Josephson oscillations are averaged over, then states of any energy count as equilibrium in this framework. We therefore argue that the excitations of a two-mode BH system can indeed provide a simple model for heat, if they are involved in a dynamical process that occurs slowly compared to their Josephson frequency. To ensure this, we allow for thermal contact between the subsystems through an additional Josephson tunneling with a much smaller coupling coefficient than that of intra-subsystem tunneling. Therefore, on the one hand the two subsystems can exchange particles on a longer time scale, on the other hand they can also exchange heat, as the amplitudes of their fast Josephson oscillations beat slowly back and forth. In fact we could show that the beating Josephson excitations do behave very similarly to heat in a peculiar way, namely second sound. This new branch in the excitation spectrum is usually associated with temperature oscillations. In the case of our four-mode system, we identify a nonlinear collective effect that makes Josephson oscillations of heat—'second Josephson oscillations'—the slowest collective excitation about an excited equilibrium state. Our analytical derivation of this effect shows, that a simple quantum mechanical Hamiltonian can yield a candidate minimalist model for the exchange of heat and further indicates how thermodynamics may emerge from quantum dynamics in small systems.

2 The two-mode system revisited

Let us at first have a look at the two-mode BH model, which has been extensively studied in many excellent contributions [9]. Our system consists of two pairs of bosonic modes each described by the standard BH Hamiltonian

$$\hat{H}_\alpha = -\frac{\Omega}{2}\left(\hat{a}_{1\alpha}^\dagger \hat{a}_{2\alpha} + \hat{a}_{2\alpha}^\dagger \hat{a}_{1\alpha}\right) + U\sum_{i=1}^{2}\hat{a}_{i\alpha}^{\dagger 2}\hat{a}_{i\alpha}^2 \tag{1}$$

where $\alpha = L, R$ labels the subsystem. To go accurately beyond the linear approximation within the standard symmetry-breaking Bogoliubov theory requires careful treatment of zero modes. However, this can easily be circumvented by following a number-conserving approach [10, 11]. Applying the Holstein-Primakoff transformation (HPT) [12], we re-write the Hamiltonian in terms of the total atom number on each side $\hat{N}_\alpha = \sum_{i=1,2}\hat{a}_{i\alpha}^\dagger \hat{a}_{i\alpha}$, and atom-moving operators $\hat{a}_\alpha^\dagger, \hat{a}_\alpha$ that commute with $\hat{N}_\alpha$:

$$\hat{a}_\alpha \equiv (\hat{a}_{+\alpha}^\dagger \hat{a}_{+\alpha})^{-1/2}\hat{a}_{+\alpha}^\dagger \hat{a}_{-\alpha}. \tag{2}$$

It can easily be shown that these operators fulfill the usual bosonic commutation relations, $[\hat{a}_\alpha, \hat{a}_\alpha^\dagger] = 1$. In terms of these operators the subsystems' Hamiltonian reads

$$\hat{H}_\alpha = -\frac{\Omega}{2}\hat{N}_\alpha + \Omega\hat{a}_\alpha^\dagger \hat{a}_\alpha + \frac{U}{2}\hat{N}_\alpha(\hat{N}_\alpha - 2) + \frac{U}{2}\hat{N}_\alpha(\hat{a}_\alpha^\dagger + \hat{a}_\alpha)^2$$
$$-\frac{U}{4}\{\hat{a}_\alpha^\dagger + \hat{a}_\alpha, \hat{a}_\alpha^\dagger \hat{a}_\alpha \hat{a}_\alpha + \hat{a}_\alpha^\dagger \hat{a}_\alpha^\dagger \hat{a}_\alpha\} + \mathcal{O}(U\hat{N}_\alpha^{-1}). \tag{3}$$

The Bogoliubov formalism is then implemented simply by defining $\hat{a}_\alpha = u_\alpha \hat{b}_\alpha + v_\alpha \hat{b}_\alpha^\dagger$ and choosing u_α, v_α to diagonalize the quadratic terms in $\hat{H}_\alpha$ while still demanding $[\hat{b}_\alpha, \hat{b}_\alpha^\dagger] = 1$. Since we are interested in dynamics slow compared to Ω, which sets the time-scale of the fast excitations of the subsystems, and in Josephson oscillations of finite but not extreme amplitude, we will apply the rotating wave approximation (RWA) by dropping all terms that do not commute with the leading term proportional to $\hat{b}_\alpha^\dagger \hat{b}_\alpha$. If we for simplicity assume the regime of large occupations, we may omit corrections of order $U\hat{N}_\alpha^{-1}$, and finally obtain

$$\hat{H}_\alpha = -\frac{\Omega}{2}\hat{N}_\alpha + \frac{U}{2}\hat{N}_\alpha(\hat{N}_\alpha - 2) + \tilde{\Omega}\hat{b}_\alpha^\dagger \hat{b}_\alpha + U_J\hat{b}_\alpha^{\dagger 2}\hat{b}_\alpha^2, \tag{4}$$

where $\tilde{\Omega} = \sqrt{\Omega(\Omega + UN)}$ is the familiar Josephson frequency of the excitations created by $\hat{b}^{\dagger}_{\alpha}$, which we call 'josons' in the following. The interaction coefficient

$$U_J \equiv -\frac{U}{4}\frac{4\Omega + UN}{\Omega + UN} \tag{5}$$

is in general of the same order as U, but has the opposite sign. In quasiparticle terms, we may say that where atoms repel each other, josons attract; and *vice versa*.

3 Four-mode Bose-Hubbard

Our complete system now is a rather symmetric four-mode BH model. In addition to the two-mode subsystems analogous to (1) we introduce a tunneling coupling between the two of them, but with a by far smaller rate coefficient $\omega \ll \Omega$ to take care of a clear time-scale separation. The full system's Hamiltonian then reads

$$\hat{H} = \hat{H}_L + \hat{H}_R - \frac{\omega}{2}\left(\hat{a}^{\dagger}_{1L}\hat{a}_{1R} + \hat{a}^{\dagger}_{2L}\hat{a}_{2R} + \text{H. c.}\right), \tag{6}$$

where the parts of the left and right subsystem, $\alpha = L, R$, are given by the two-mode Hamiltonian (1). Since $\hat{H}$ conserves the total atom number $\hat{N} = \sum_{i,\alpha}\hat{a}^{\dagger}_{i\alpha}\hat{a}_{i\alpha}$ we are left with the three nontrivial tunneling modes. The frequencies of these elementary excitations above the N-atom ground state can easily be computed in linear Bogoliubov approximation:

$$\tilde{\omega} = \sqrt{\omega(\omega + UN)}, \quad \tilde{\Omega} = \sqrt{\Omega(\Omega + UN)} \text{ and}$$
$$\tilde{\Omega}' = \sqrt{(\Omega + \omega)(\Omega + \omega + UN)}. \tag{7}$$

In the $\omega \ll \Omega$ case we consider here, only one of these three elementary modes, $\tilde{\omega}$, is slow, the others are fast. And although the excitations of the Bogoliubov modes provide a complete basis for the N-particle many-body Hilbert space, there still exists in fact a second distinct collective mode with a low frequency of order ω in this system, analogously to e. g. second sound in superfluid helium [13] or even ordinary sound in air. The analogy, however, to second sound in helium II is remarkable, since this effect is understood as the decoupling of temperature and density waves. That leads to a second branch in the dispersion relation corresponding to temperature waves. This rather unusual oscillating behaviour of heat is exactly what is mimicked by our simple model. The interference beat between the two nearly degenerate high-frequency modes provides the extra collective mode with the low frequency

$$\omega_J \equiv (\omega/\tilde{\Omega})(\Omega + UN/2) = \tilde{\Omega}' - \tilde{\Omega} + \mathcal{O}(\omega^2/\Omega). \tag{8}$$

In fact it can be shown, that to leading order in the small frequency ratio ω/Ω, but nonlinearly in excitation amplitude, within the low-frequency regime this beat mode is an independent excitation in its own right, whose frequency is actually shifted away from ω_J by nonlinear effects.

4 Second Josephson oscillations

To see this, we at first perform a similar HPT for our two atom numbers, while assuming that our system is in an eigenstate of the conserved total $\hat{N}_L + \hat{N}_R$, with eigenvalue N. This defines a number-conserving operator $\hat{a}$, which satisfies $[\hat{a}, \hat{a}^\dagger] = 1$ and transfers atoms between the L and R subsystems. We can therefore re-write our full Hamiltonian (6) in terms of $\hat{b}_{L,R}$, $\hat{a}$, and N. To leading order in ω/Ω, our entire Hamiltonian then simply is $\Omega\hat{J}$, for

$$\hat{J} \equiv \hat{b}_L^\dagger \hat{b}_L + \hat{b}_R^\dagger \hat{b}_R. \tag{9}$$

In classical terms, the 'total joson number' $\hat{J}$ is an adiabatic invariant and so we may let $\hat{J} \to J$ in the low-frequency theory. By means of a RWA we can compute low frequency dynamics nonlinearly in excitation amplitude, but to leading order in small frequency ratios, by dropping all terms in $\hat{H}$ that do not commute with $\hat{J}$:

$$\hat{H} = \omega\hat{a}^\dagger\hat{a} + \frac{U}{4}N(\hat{a}^\dagger + \hat{a})^2 - \frac{\omega_J}{2}\left(\hat{b}_L^\dagger\hat{b}_R + \hat{b}_R^\dagger\hat{b}_L\right) + \frac{U_J}{2}\sum_{\alpha=L,R}\hat{b}_\alpha^\dagger\hat{b}_\alpha^\dagger\hat{b}_\alpha\hat{b}_\alpha$$

$$+\frac{U}{2}\sqrt{\frac{\Omega N}{\Omega + UN}}(\hat{a}^\dagger + \hat{a})(\hat{b}_L^\dagger\hat{b}_L - \hat{b}_R^\dagger\hat{b}_R) \tag{10}$$

We can now perform a final HPT to express $\hat{b}_\alpha$ in terms of $\hat{J}$ and a J-conserving operator $\hat{b}$ that transfers josons between subsystems. We then linearize in $\hat{a}$ and $\hat{b}$. This linearization, however, is different from the standard Bogoliubov approximation of retaining only quadratic terms in fluctuations of the original $\hat{a}_{i\alpha}$ about the mean-field ground state. Via the HPT certain dominant nonlinear interactions among atoms have been included in the linearized interactions among josons. One may also say that we do pursue the standard Bogoliubov approach, but applied to a two-mode system of interacting conserved atoms and josons. In this sense the extra low-frequency mode is a non-elementary collective mode in the full theory, but an elementary excitation within the adiabatic effective theory for low frequency dynamics. The linearized Hamiltonian reads

$$\hat{H} = \omega\hat{a}^\dagger\hat{a} + \frac{U}{4}N(\hat{a}^\dagger + \hat{a})^2 + \omega_J\hat{b}^\dagger\hat{b} + \frac{U_J}{4}J(\hat{b}^\dagger + \hat{b})^2$$

$$+ \frac{U}{2}\frac{\Omega}{\tilde{\Omega}}(NJ)^{1/2}(\hat{a}^\dagger + \hat{a})(\hat{b}^\dagger + \hat{b}). \tag{11}$$

In the first line we recognize two different forms of the standard single Josephson junction Hamiltonian. One provides Josephson oscillations of atoms between L and R subsystems, with Josephson frequency $\tilde{\omega}$, while the second implies *Josephson oscillations of josons* with the new frequency

$$\tilde{\omega}_J = \sqrt{\omega_J(\omega_J + U_J J)}. \tag{12}$$

The two Josephson modes are exactly analogous, except that U_J and U are of opposite sign; the coupling of the atom and joson mode arises because the frequency of fast Josephson oscillations in each subsystem depends on the respective atom number. The decoupled collective mode frequencies in (11) finally yield two slow elementary excitation frequencies for fixed N and J:

$$\tilde{\omega}_\pm^2 = \frac{\tilde{\omega}^2 + \tilde{\omega}_J^2}{2} \pm \left[\left(\frac{\tilde{\omega}^2 - \tilde{\omega}_J^2}{2} \right)^2 + \frac{\omega\omega_J \Omega U^2 N J}{\Omega + UN} \right]^{1/2}. \tag{13}$$

In the low-amplitude regime $J \ll N$ the low atom Josephson frequency $\tilde{\omega}_+ \approx \tilde{\omega}$ is always essentially unchanged from the standard Bogoliubov theory, but it is joined by a second Josephson oscillation whose frequency $\tilde{\omega}_-$ can be significantly lower (for $U > 0$), because U_J is negative.

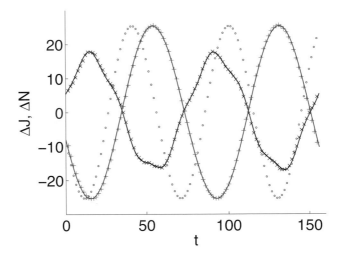

Fig. 2. Time evolution of $\Delta J = b_L^\dagger b_L - b_R^\dagger b_R$ and ΔN in the mean-field system for the parameters $\Omega = 1$, $\omega = 0.1$, $UN = 1$, $N = 10^4$ and $J = 253$. Red +: numerical mean-field ΔJ. Red solid line: oscillation at second Josephson frequency $\tilde{\omega}_-$ computed in the text, fitted for phase. Black ○: oscillation at beat-frequency ω_J. Blue ×: numerical ΔN. Blue solid line: oscillation at a beat of the frequencies $\tilde{\omega}_+$ and $\tilde{\omega}_-$ fitted for amplitude and phase.

To test our theory of second Josephson oscillations numerically, we solve the four-mode Gross-Pitaevskii nonlinear Schrödinger equation for the classical limit of our system. We start with the N-particle mean-field ground state, then excite josons by adding a time-dependent potential tilt to each subsystem, which varies periodically with the Josephson frequency $\tilde{\Omega}$. To excite the second Josephson mode we give the tilt drives different amplitudes for the L and R pairs, so that slightly more josons are excited on one side than on the other. The resulting evolution, after the drive is turned off, is shown in Fig. 2 for a typical case. The agreement with our linearized theory is excellent, and the nonlinear shifting of the second Josephson frequency well below the trivial Bogoliubov beat is clear. Some atom population oscillation between L and R subsystems occurs at frequency $\tilde{\omega}_+$, this mode having been slightly excited by our driving procedure, but the main atom oscillation is exactly opposite in phase to the joson oscillation. We note that while in incompressible superfluids second sound includes no significant particle density change, the coupling of density and heat oscillations is expected for second sound in compressible superfluids [14]. The coupling between the slow atom and joson oscillations can be interpreted as a simple example of the interconversion of heat and work.

5 Discussion

Taking advantage of the small ratio between the low and high tunneling rates, we are able to analyze collective excitations in this model beyond the linear Bogoliubov approximation and show, that the low-frequency behavior of the joint system consists of two Josephson oscillations: the elementary Bogoliubov excitation whereby the two subsystems exchange atoms with each other, and a nonlinear beat mode by which they exchange josons, i. e. heat. Insofar as a Josephson oscillation represents a two-mode analog of Bogoliubov zero sound in a dilute Bose gas, we identify our second Josephson mode as an analog of second sound. We offer it as a simple example of a typical phenomenon of mesoscopic thermodynamics. In our simple model, however, Second Law phenomenology, irreversible diffusion of heat, cannot be observed and may require larger system sizes. Thus, looking beyond our simple model is necessary to investigate the emergence of irreversibility. Classical discussions of thermalization usually suggest, at least implicitly, that chaotic behaviour is involved. Especially the question if quantum chaos plays a role in this process deserves extensive further studies.

References

1. J. M. Deutsch: Phys. Rev. A **43**, 2046 (1991).
2. M. Srednicki: Phys. Rev. E **50**, 888 (1994).
3. T. Kinoshita, T. Wenger, D. S. Weiss: Nature **440**, 900 (2006).
4. M. Rigol, V. Dunjko, M. Olshanii: Nature **452**, 854 (2008).
5. M. Eckstein, M. Kollar, P. Werner Phys. Rev. Lett. **103**, 056403 (2009).
6. M. Rigol: Phys. Rev. Lett. **103**, 100403 (2009).
7. L. D. Landau, E. M. Lifshitz: *Statistical Physics, Part 1* (Pergamon Press, Oxford, 1980), Chap. 1, §1 .
8. M. P. Strzys, J. R. Anglin: Phys. Rev. A **81**, 043616 (2010).

9. G. J. Milburn, J. Corney, E. M. Wright, D. F. Walls: Phys. Rev. A **55**, 4318 (1997).
10. C. W. Gardiner: Phys. Rev. A **56**, 1414 (1997).
11. Y. Castin, R. Dum: Phys. Rev. A **57**, 3008 (1998).
12. T. Holstein, H. Primakoff: Phys. Rev. **58**, 1098 (1949).
13. P. Nozières, D. Pines: *The Theory of Quantum Liquids* (Addison-Wesley, Redwood City, 1990), Vol. II.
14. A. Griffin, E. Zaremba: Phys. Rev. A **56**, 4839 (1997).

Quantum statistical synchronization of non-interacting particles

Malte C. Tichy[1], Markus Tiersch[1,2], Fernando de Melo[1,3], Florian Mintert[1,4], and Andreas Buchleitner[1]

[1] Physikalisches Institut, Albert–Ludwigs–Universität Freiburg, Hermann–Herder–Strasse 3, D–79104 Freiburg, Germany
[2] Institute for Quantum Optics and Quantum Information, Austrian Academy of Sciences, Technikerstrasse 21A, A–6020 Innsbruck, Austria
[3] Instituut voor Theoretische Fysica, Katholieke Universiteit Leuven, Celestijnenlaan 200D, B–3001 Heverlee, Belgium
[4] Freiburg Institute for Advanced Studies, Albert-Ludwigs-Universität Freiburg, Albertstrasse 19, D-79104 Freiburg, Germany

Abstract. A full treatment for the scattering of an arbitrary number of bosons through a Bell multiport beam splitter is presented that includes all possible output arrangements. Due to exchange symmetry, the event statistics differs dramatically from the classical case in which the realization probabilities are given by combinatorics. A law for the suppression of output configurations is derived and shown to apply for the majority of all possible arrangements. Such multiparticle interference effects dominate at the level of single transition amplitudes, while a generic bosonic signature can be observed when the average number of occupied ports or the typical number of particles per port is considered. The results allow to classify in a common approach several recent experiments and theoretical studies and disclose many accessible quantum statistical effects involving many particles.

1 Introduction

Non-interacting, distinguishable particles exhibit independent and therefore uncorrelated behavior. Due to the bosonic or fermionic nature of identical particles, however, such statement is no longer true for indistinguishable particles, even if no interaction takes place. For example, the bosonic nature of photons is impressively demonstrated by their statistical behavior in a Hong-Ou-Mandel (HOM) setup [1]. Here, two identical photons are sent simultaneously (within their coherence time) through the two input ports of a balanced beam splitter. Due to the lack of interaction between the photons, one would not expect any correlations in the number of photons measured at both output ports. For fully indistinguishable photons, however, the particles always leave the setup together, and never exit at different ports.

Such synchronization of two non-interacting particles has lead to many applications in quantum information sciences. The visibility of the HOM-dip quantifies the indistinguishability of two photons [2]. Thereby, the quality of single-photon sources can be tested [3]. The maximally entangled $\rangle\Psi^-$ Bell-state (in any degree of freedom carried by the photons) can be detected or created, since it leads to an unambiguous signature in the setup [4]. This projection onto an entangled state can be applied in entanglement swapping protocols [5] and quantum metrology [6].

It is therefore of great interest to generalize the HOM setup for more than two photons and more than two input or output ports, i.e. to n particles that are scattered in a setup with n input and output ports. This would allow applications such as entanglement swapping or entanglement detection for many particles and the experimentally controlled transition from indistinguishability to distinguishability for many identical particles [7, 8].

While a comprehensive understanding of this scattering scenario is not yet available due to the complexity of the problem and the prohibitive scaling of the number of output states, several steps have been undertaken in this direction. The measurement of the enhancement of events with all particles in one port - bunching events, was realized experimentally [9, 10], a prediction for the suppression of coincident events for a specially designed biased setup with three particles and three input ports was presented [11]. The case of a Bell multiport beam splitter [12, 13] which redistributes n incoming particles to n ports in an unbiased way was discussed in [14], where it was shown that coincident events are suppressed when n is even.

In this contribution, we extend our recent results [15] on the characterization of the probabilities of *all* possible output events of the Bell multiport beam splitter when n particles are prepared in the n input ports. Such treatment enables a general understanding of multiparticle interference effects, as well as on the average behavior of bosons. It hence unifies previous experimental and theoretical work on multiport beam splitters, and opens up new perspectives for the experimental verification and exploitation of bosonic multiparticle behavior.

2 Formalism

2.1 Setup and notation

We consider a scattering scenario in which n particles are initially evenly distributed among n input modes. They are scattered by the multiport beam splitter and exit among n output ports. The probability for each particle to exit at any port is the same, i.e. $1/n$. Such setup can be realized for photons via a simple beam splitter in the two-photon HOM case, as illustrated in Fig. (1a). A pyramidal combination of beam splitters with different reflection/transmission rates yields the generalization for n ports and particles, such setup is shown in Fig. (1b), for $n = 5$.

(a)   (b)

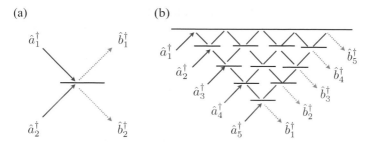

Fig. 1. (a) Two-photon HOM scattering setup with two photons impinging on a balanced beam splitter. (b) Generalization: multiport beam splitter, here with five incoming and outgoing modes.

We denote arrangements of n particles in the n modes by a vector $\boldsymbol{s} = (s_1, s_2, \ldots s_n)$, with s_k the number of particles in the output mode k, and $\sum_{i=1}^{n} s_i = n$. Alternatively, we define the *port assignment vector* $\boldsymbol{d}$ of length n with entries that specify each particle's output port. It is constructed by concatenating s_j times the port number j:

$$\boldsymbol{d} = \oplus_{j=1}^{n} \oplus_{k=1}^{s_j} (j), \tag{1}$$

e.g., for the arrangement $\boldsymbol{s} = (2, 1, 0, 2, 0)$, we find $\boldsymbol{d}(\boldsymbol{s}) = (1, 1, 2, 4, 4)$.

2.2 Distinguishable particles

For distinguishable particles, no many-particle interference takes place, and the probability for a certain arrangement $\boldsymbol{s}$ is given by simple combinatorics:

$$P_{\text{class}}(\boldsymbol{s}) = \frac{1}{n^n} \frac{n!}{\prod_{j=1}^{n} s_j!} \tag{2}$$

Due to the lack of interference phenomena, we call this situation "classical". In accordance to our intuition, probabilities are summed instead of amplitudes. Two classes of events are especially interesting due to their extremal character. *Coincident events*, i.e. $\boldsymbol{s}_c = (1, 1, \ldots 1)$, are realized with probability $n!/n^n$. *Bunching events*, with all particles at one output mode k, correspond to $s_k = n$ and thus to $\boldsymbol{s}_b = (0, 0, .., n, ..0)$. They are realized with probability $1/n^n$ and suppressed by a factor of $n!$ with respect to the coincident events. For large n, both events are highly unlikely, extreme cases.

2.3 Indistinguishable bosons

We reformulate the scattering scenario for identical particles, in second quantization. Applications of our study are feasible with today's optical technologies [12], therefore, we focus on bosons. The initial state with one particle in each input port reads $|\Psi\rangle = \prod_{i=1}^{n} \hat{a}_i^\dagger |0\rangle$. The single-particle unitary evolution induced by the scattering setup acts on all particles independently and maps the input port creation operators $\hat{a}_i^\dagger$ to output creation operators $\hat{b}_i^\dagger$ via a unitary matrix U [16], such that $\hat{b}_j^\dagger = \sum_{k=1}^{n} U_{jk} \hat{a}_k^\dagger$. The unbiased Bell multiport beam splitter under consideration here corresponds to the unitary operation given by the Fourier matrix, defined for any dimension n by

$$U_{jk} = \frac{e^{\frac{2\pi i}{n}(j-1)(k-1)}}{\sqrt{n}}. \tag{3}$$

The possible states with fixed particle number per port after the scattering process read

$$|\Phi(\boldsymbol{s})\rangle = \left(\prod_{j=1}^{n} \frac{1}{\sqrt{s_j!}} \left(\hat{b}_j^\dagger \right)^{s_j} \right) |0\rangle. \tag{4}$$

The transition probability to a specific output arrangement $\boldsymbol{s}$ can be written with the help of the port assignment vector (Eq. 1) as

$$P_{\mathrm{qm}}(\boldsymbol{s}) = |\langle \Psi | \Phi(\boldsymbol{s}) \rangle|^2 = \frac{1}{\prod_j s_j!} \left| \sum_{\sigma \in P_n} \prod_{j=1}^{n} U_{d_j(\boldsymbol{s}), \sigma(j)} \right|^2, \tag{5}$$

where P_n denotes the set of all permutations of $\{1, .., n\}$. This coherent sum over $n!$ terms expresses the interference that occurs between all many-particle amplitudes that lead to the same output state.

2.4 Equivalence classes

In order to discuss the behavior of the scattering system, it is necessary to identify classes of final states that occur with equal probability, within the quantum and the classical case. In the latter case, the realization probability of any arrangement $\boldsymbol{s}$, Eq. (2), remains invariant under permutation of the output ports s_k. Hence we can define *classical* equivalence classes which identify arrangements related to each other by permutation. Ultimately, however, all final events $\boldsymbol{s}$ have to be considered as inequivalent. In the case that the scattering matrix is a Fourier matrix such as given in Eq. (3), some symmetry properties allow to reduce considerably the number of equivalence classes. One indeed finds that the amplitude (5) is invariant under cyclic and anticyclic permutations. This allows us to define a *quantum* equivalence relation between arrangements, and N_{quant} associated quantum equivalence classes.

The number of classical equivalence classes corresponds to the partition number, i.e. the number of possibilities to write an integer as sum of positive integers, while the total number of inequivalent events grows much faster, it is given by $N_{\mathrm{total}} = \frac{(2n)!}{2(n!)^2}$. For comparison, the number of equivalence classes are given in Table 1.

3 Event-suppression law

In general, the evaluation of the transition probabilities in Eq. (5) is a difficult task and cannot be performed in polynomial time with n [17]. It is, however, possible to exploit the symmetry of the matrix U to formulate a powerful law which predicts the suppression of final events. Indeed, since only n-th roots of unity appear in the Fourier matrix (Eq. 3), also every term of the $n!$ summands in Eq. (5) can be written as such. Thereby, Eq. (5) turns into

$$\langle \Psi | \Phi(\boldsymbol{s}) \rangle = \sum_{k=0}^{n-1} c_k e^{i\frac{2\pi}{n}k}, \tag{6}$$

where the c_k are natural numbers which give the cardinality of the following sets, defined in analogy to [18],

$$u_r(\boldsymbol{s}) = \left\{ \sigma \,\middle|\, \Theta_{n,\boldsymbol{s}}(\sigma) \equiv \sum_{l=1}^{n} d(\boldsymbol{s})_l \sigma(l) = r \bmod n \right\}, \tag{7}$$

with $c_r = |u_r(\boldsymbol{s})|$. The sum corresponds to the position of the barycenter of the set of points $\{c_k e^{i\frac{2\pi}{n}k} | k \in \{1, .., n\}\}$ in the complex plane. We set $Q = \bmod(\sum_{l=1}^{n} d_l(\boldsymbol{s}), n)$, and define an operation γ which acts on permutations such that $\gamma(\sigma)(k) = \sigma(k) + 1 \bmod n$. We find that $\Theta_{n,\boldsymbol{s}}(\gamma(\sigma)) = \Theta_{n,\boldsymbol{s}}(\sigma) + Q$. Thus, if $Q \neq 0$, the repeated application of γ gives us a bijection between all pairs of $u_{r+a \cdot Q}$, for $a \in \{0, 1, .., n-1\}$. Hence, we find

$$\forall r \in \{0, .., n-1\}, \forall a \in \mathbb{N} : c_{r+a \cdot Q} = c_r. \tag{8}$$

If $Q \neq 0$, the set of points $\{c_k e^{i\frac{2\pi}{n}k} | k \in \{1, .., n\}\}$ describes several interlaced polygons centered at the origin, ensuring that the sum vanishes, hence the process with the final state $\boldsymbol{s}$ is suppressed in this case. Thus, without knowing the values of the individual c_k, and only by symmetry properties, it is possible to predict that the total sum (Eq. 5) vanishes. This observation allows us to formulate:

$$Q(\boldsymbol{s}) := \mathrm{Mod}\left(\sum_{l=1}^{n} d_l(\boldsymbol{s}), n \right) \neq 0 \;\Rightarrow\; \langle \Psi | \Phi(\boldsymbol{s}) \rangle = 0. \tag{9}$$

The law can be applied on any final state in an efficient way: consider, e.g., $n = 6$ and $\boldsymbol{s}_1 = (2, 1, 2, 1, 0, 0)$. The port assignment vector reads $(1, 1, 2, 3, 3, 4)$, and one finds $Q(\boldsymbol{s}_1) = 2$, and this event is hence strictly suppressed. Unexpectedly though, the event $\boldsymbol{s}_2 = (0, 1, 2, 0, 2, 1)$, which is obtained from $\boldsymbol{s}_1$ by simple permutation, gives $Q(\boldsymbol{s}_2) = 0$ due to the different port assignment vector $(2, 3, 3, 5, 5, 6)$. It is actually enhanced by a factor larger than seven as compared to the classical event probability (also see Table 2).

n	N_{total}	N_{class}	N_{quantum}	N_{law}	N_{supp}
2	3	2	2	1	0
3	10	3	3	1	0
4	35	5	8	5	0
5	126	7	16	10	0
6	462	11	50	38	2
7	1716	15	133	105	0
8	6435	22	440	371	0
9	24310	30	1387	1201	0
10	92378	42	4752	4226	96
11	352716	56	16159	14575	0
12	1352078	77	56822	51890	1133
13	5200300	101	200474	184626	0
14	20058300	135	718146	666114	2403

Table 1. Total number of events (N_{total}), classical equivalence classes (N_{class}), quantum equivalence classes (N_{quantum}), classes that therewithin fulfill the law 9 (N_{law}), and suppressed classes which are not predicted by Eq. 9 (N_{supp}).

n	s	Enhancement
3	(003)	6
	(111)	3/2
4	(0004)	24
	(0202) , (0121)	8/9
5	(00005)	120
	(00131), (01103)	15/2
	(00212), (01022)	10/3
	(11111)	5/24
6	(000006)	720
	(002004), (000141), (010104), (000303), (001032), (000222)	144/5
	(020202), (001113), (012021)	36/5

Table 2. Nonsuppressed output states, together with the corresponding quantum enhancement, i.e., the ratio of quantum to classical event probability.

3.1 Suppressed arrangements

It is possible to estimate the number of suppressed arrangements predicted with the help of (9) by a simple argument. Since the number of arrangement N_{quant} is much larger than n, we can assume that the $Q(s)$ are uniformly distributed in the interval $[0, \ldots, n-1]$ for the ensemble of events s. Then the probability to find a suppressed arrangement is given by the weight of nonvanishing values of $Q(s)$, i.e., by $(n-1)/n = 1 - 1/n$. This estimate is also numerically confirmed in the values shown in Table 1.

3.2 Application of the suppression law

For $n = 2..6$, we list the unsuppressed arrangements in Table 2, together with their *quantum enhancement*, i.e., the ratio of quantum-to-classical event probability. From the results, it is apparent that the behavior of the system becomes more extreme, the more particles are involved: the enhancement factor is bounded from above by $n!$, a value that is reached for the enhancement for bunching events with $s = (n, 0, .., 0)$. For the 50 quantum equivalent arrangements that exist for $n = 6$, we show the classical and quantum probabilities in Fig. 2. Two arrangements are suppressed although they do not fulfill the requirements of the law: $s = (0, 1, 2, 1, 0, 2)$ and $s = (0, 1, 1, 2, 1, 1)$. In Fig. 3, we show the values of the corresponding c_k in the complex plane. One can easily see that while the values do not lie on polygons, the sum of all contributions still vanishes. Such situations are exceptional, as can be seen from Table 1.

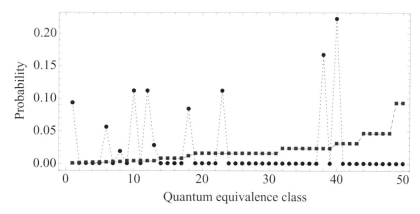

Fig. 2. Quantum (blue circles) and classical (red squares) probability for the realization of the 50 different quantum equivalence classes for $n = 6$ particles, sorted by their classical realization probability. Note that most states are fully suppressed in the quantum case, only a few are highly enhanced with respect to the classical case.

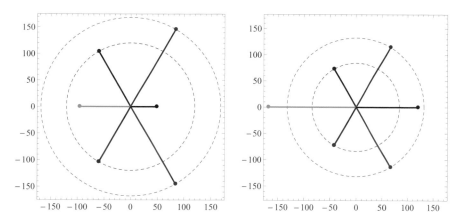

Fig. 3. Illustration of the corresponding c_k in the complex plane for the process with $s = (0, 1, 2, 1, 0, 2)$ (left hand side) and $s = (0, 1, 1, 2, 1, 1)$ (right hand side). The sum of the complex numbers vanishes while the points do not lie all on circles.

4 Bosonic behavior

As we have seen in the last section, the implications of Eq. (9) on the realization probabilities of single events are important for the overall behavior of the system: most events are totally suppressed, while only few remain which are highly enhanced. Intuitively, one would expect that events with many particles in one port are generally favored by bosons. Indeed, bunching events are always enhanced by a factor of $n!$ with respect to the classical case. The number of particles in one port or the number of occupied ports does, however, turn out not to be a good indicator for the enhancement or the suppression of a certain event. For example, events of the type $s = (n - 1, 1, 0, .., 0)$ could be expected to be enhanced due to the bosonic nature of the particles, while they actually turn out to be strictly suppressed, for all n. Thus, at the level of the event probabilities of single arrangements, interference effects dominate, and the bosonic nature of the particles is not apparent at all.

Such general bosonic behavior is recovered when a coarse-grained grouping of many final arrangements in larger classes is performed. Such classes can be characterized, e.g., by the number of occupied ports k, by the number m of particles in one port, or by the classical equivalence classes. The event probability for such a class is given by the sum of the probabilities of the single events that pertain to the class.

When performing such average, we expect that interference effects disappear while the bosonic enhancement of states with many particles in one port persists. This can be also seen in our formalism: according to (5), the probabilities $P_{qm}(s)$ are given in terms of a sum over permutations of *scattering amplitudes*, i.e., over complex numbers of equal modulus (products of matrix-elements of U_{jk}). Since these numbers typically have different phases, they tend to add up destructively. However, all $s_j!$ permutations σ that interchange the s_j particles that exit in port j leave the scattering amplitudes invariant, so that $s_j!$ terms in the sum have equal phases and add up constructively. This motivates the following approximation for the transition probability (5):

$$P_{\text{approx}}(s) = \frac{\left(\prod_j s_j! \right) P_{\text{class}}(s)}{\sum_r \left(\prod_j r_j! \right) P_{\text{class}}(r)}. \tag{10}$$

We show the probability distribution for the number of occupied ports, for the classical calculation (Eq. 2), for the bosonic quantum case (Eq. 5), and for our approximation (Eq. 10), for $n = 14$, in Fig. 4. As expected, bosons always tend to occupy less output ports than in the classical case. This behavior is persistent for any n. Furthermore, Fig. 4 shows that the approximation (Eq. 10) predicts the actual outcome very well for most k, and only fails for events with almost all or almost no sites occupied. This is easily understood, since, for very small or very large k, few distinct equivalence classes contribute to these event groups. Then, again interference dominates the event probability, rather than bosonic behavior.

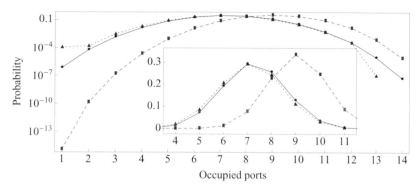

Fig. 4. Event probability for a given number of occupied ports, for $n = 14$. Red squares denote classical combinatorics, blue triangles the quantum calculation, and black circles our (bosonic) estimate for the quantum result. The inset shows the same distribution on a linear scale. Events with 14 occupied ports are strictly suppressed in the quantum case.

Also the event probability for a given number of particles in one single port is well described by our estimate (Eq. 10). For 14 particles, the probability distribution is shown in Fig. 5. Again, we see a dramatic difference between the classical and quantum case, especially for the probability to find a large number of particles in one port.

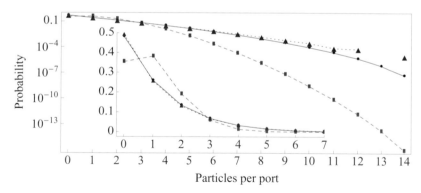

Fig. 5. Probability to find exactly k particles (horizontal axis) in one port, for $n = 14$. Red squares denote the classical, blue triangles the quantum calculation, and black circles the estimate (10). The inset shows the distribution on a linear scale for small k. Events with 13 particles in one port are totally suppressed for bosons.

A further application of the bosonic approximation can be performed when we group the 718146 events according to their 135 classical equivalence classes. The resulting probabilities are shown in Fig. 6. While this grouping is much more fine than in the previous two examples, the difference between the clas-

sical and quantum case is still very well pronounced and reproduced by the
estimate.

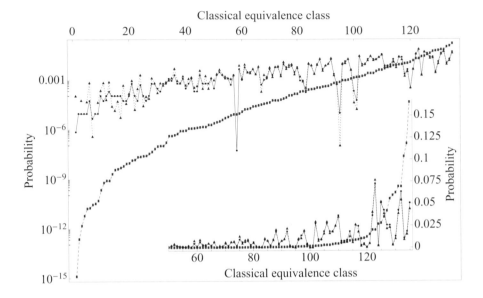

Fig. 6. Quantum (Blue triangles) and classical probability (Red rectangles) and
estimate (Black circles) for the realization of events grouped according to the 135
classical equivalence classes. The classical equivalence classes are ordered according
the classical realization probability given by Eq. (2). The classical probability is
therefore – by construction – a monotonically increasing function. The inset shows
a detail of the probability distribution in linear scale.

5 Conclusions and outlook

Synchronization of non-interacting particles might seem contradictory by construction. However, it turns out to be possible due to the exploitation of quantum statistical effects stemming from the indistinguishability of particles. We
generalized the most prominent example for such a behavior, the HOM effect,
to n particles and n ports on two different levels: interference effects inhibit
the realization of most possible events for single transition amplitudes, while
general statistical characteristics with smooth bosonic behavior emerge which
are efficiently approximated by Eq. (10). On the fine as well as on the coarse
grained scale, however, quantum and classical transmission probabilities differ
dramatically.

Acknowledgements

M.C.T. acknowledges financial support by Studienstiftung des deutschen Volkes, F.d.M. by the Belgium Interuniversity Attraction Poles Programme P6/02, and F.M. by DFG grant MI 1345/2-1, respectively.

References

1. C. K. Hong, Z. Y. Ou, L. Mandel: Phys. Rev. Lett. **59**, 2044 (1987).
2. Z. Y. Ou: Phys. Rev. A **74**, 063808 (2006).
3. F. W. Sun, C. W. Wong: Phys. Rev. A **79**, 013824 (2009).
4. Y. H. Shih, C. O. Alley: Phys. Rev. Lett. **61**, 2921 (1988).
5. M. Halder, A. Beveratos, N. Gisin et al: Nature Physics **3**, 692 (2007).
6. P. Walther, J.-W. Pan, M. Aspelmeyer et al: Nature **429**, 6988 (2004).
7. M. C. Tichy, F. de Melo, M. Kuś et al: Entanglement of identical particles and the detection process. Preprint arXiv:0902.1684.
8. M. C. Tichy, H.-T. Lim, Y.-S. Ra et al: Four-Photon Distinguishability Transition. Preprint arXiv:1009.4998.
9. Z. Y. Ou, J.-K. Rhee, L. J. Wang: Phys. Rev. Lett. **83**, 959 (1999).
10. X.-L. Niu, Y.-X. Gong, B.-H. Liu et al: Opt. Lett. **34**, 1297 (2009).
11. R. A. Campos: Phys. Rev. A **62**, 013809 (2000).
12. M. Żukowski, A. Zeilinger, M. A. Horne: Phys. Rev. A **55**, 2564 (1997).
13. A. Vourdas, J. A. Dunningham: Phys. Rev. A **71**, 013809 (2005).
14. Y. L. Lim, A. Beige: N. J. Phys. **7**, 155 (2005).
15. M. C. Tichy, M. Tiersch, F. de Melo et al: Phys. Rev. Lett. **104**, 220405 (2010).
16. R. A. Campos, B. E. A. Saleh, M. C. Teich: Phys. Rev. A **40**, 1371 (1989).
17. Y. L. Lim, A. Beige: Phys. Rev. A **71**, 062311 (2005).
18. R. L. Graham, D. H. Lehmer: J. Austral. Math. Soc. **21**, 487 (1976).

Correlations in free fermionic systems

Mark Fannes and Jeroen Wouters

Institute of Theoretical Physics
K.U.Leuven, Belgium
mark.fannes@fys.kuleuven.be
jeroen.wouters@fys.kuleuven.be

Abstract. Correlations in quantum systems are studied in terms of conditional state spaces. Their use is illustrated for qubit Werner states. The main aim of this paper, however, is to characterize conditional state spaces in the context of free fermionic systems.

Correlations in quantum systems play an important role in practical applications such as quantum teleportation and quantum key distribution but also in more fundamental matters such as violation of Bell's inequalities. Nevertheless, the nature of quantum correlations is not fully understood, think for example of the characterization of entanglement.

In this text we propose a description of correlations in bipartite quantum systems: we study the collection of states that can be obtained on one party by perturbing the other party. We call this set the conditional state space. Through some applications, we show that the geometry of this set of states can tell us something about the correlations in the bipartite state. More applications can be found in [1].

We then move on to conditional state spaces of free fermionic states. Free fermionic states arise for example in the description of gases of non-interacting Fermions [2]. As these states are about the simplest states one can construct respecting the fermionic anti-commutation relations and as they are fully determined by their second order moments, they can be considered as fermionic gaussian states. For these states we derive two characterizations of the conditional state space: the first is in terms of inequalities for the two-point correlators, the second uses an auxiliary state space and completely positive maps that preserve the free nature of fermionic states.

1 Conditional state spaces

Conditional state spaces capture the correlations between subsystems of a composite system. Suppose that we have a bipartite state ω on a system described by two finite-dimensional observable algebras $\mathcal{A}$ and $\mathcal{B}$:

$$\omega : \mathcal{A} \otimes \mathcal{B} \to \mathbb{C}\,.$$

By fixing an operator B in $\mathcal{B}$, we get a linear functional ω_B on the algebra $\mathcal{A}$:

$$\omega_B : \mathcal{A} \to \mathbb{C} : A \mapsto \omega(A \otimes B)\,.$$

This linear functional is not necessarily positive or normalized, hence such an ω_B is not necessarily a state on $\mathcal{A}$. For this reason, we require B to be positive and impose

$$\omega_B(\mathbb{1}) = \omega(\mathbb{1} \otimes B) = 1\,.$$

We say that the positive, normalized linear functional ω_B is a state conditioned by B.

All such states form the conditional state space corresponding to ω on the party $\mathcal{A}$:

$$S_{\mathcal{A}} = \{\omega_B | B \geq 0\,,\ \omega(\mathbb{1} \otimes B) = 1\}\,.$$

Similarly, the conditional state space on the second party $\mathcal{B}$ is:

$$S_{\mathcal{B}} = \{\omega_A | A \geq 0\,,\ \omega(A \otimes \mathbb{1}) = 1\}\,.$$

These conditional state spaces play an important role in the characterization of finitely correlated states (FCS) [3]. Pure FCS are often called matrix product states (MPS) and are important in the context of ground states of quantum spin chains. General finitely correlated states on spin chains generalize the notion of Markov chains. Without entering in technical details, if $\mathcal{A}$ is the operator algebra of a site of the chain, we denote the algebra of the chain by $\mathcal{A}^{\mathbb{Z}}$. If one cuts this doubly infinite chain into two semi-infinite chains $\mathcal{A}^{\mathbb{Z}\backslash\mathbb{N}}$ and $\mathcal{A}^{\mathbb{N}}$:

$$\mathcal{A}^{\mathbb{Z}} = \mathcal{A}^{\mathbb{Z}\backslash\mathbb{N}} \otimes \mathcal{A}^{\mathbb{N}}\,,$$

then the finitely correlated states are those that have a finite-dimensional conditional state space with respect to this split, even though the algebras involved are infinite dimensional. These states can be considered as quantum Markov chains since they coincide with the classical hidden Markov chains if the observable algebras are Abelian. For more details see [3].

The geometry and the dimension of the conditional state space can inform us about the nature and the strength of the correlations between the parties. It is for example easily seen that a product state $\omega(A \otimes B) = \omega_1(A)\omega_2(B)$ leads to a trivial conditional state space.

Furthermore, conditional state spaces of separable states can be described by a classical model. As the set of separable states is convex, we can use Caratheodory's theorem to decompose a separable density matrix ρ into d separable pure states, with $d \leq d_1^2 d_2^2$, d_1 and d_2 being the dimensions of the subsystems:

$$\rho = \sum_{\alpha=1}^{d} \lambda_\alpha |\varphi_\alpha\rangle\langle\varphi_\alpha| \otimes |\psi_\alpha\rangle\langle\psi_\alpha|, \quad \sum_{\alpha=1}^{d} \lambda_\alpha = 1, \ \lambda_\alpha \geq 0$$

In this case, the density matrix of the conditional state can be written as follows:

$$\omega_B(A) = \mathrm{Tr}\rho(A \otimes B) = \sum_{\alpha=1}^{d} (\lambda_\alpha\langle\psi_\alpha|B\psi_\alpha\rangle)\langle\varphi_\alpha|A\varphi_\alpha\rangle\,.$$

As $B > 0$ and $\omega_B(\mathbb{1}) = 1$ the coefficients $(\lambda_\alpha\langle\psi_\alpha|B\psi_\alpha\rangle)$ are positive and add up to one. Hence all conditional states can be described by a convex combination of the d pure states $|\varphi_\alpha\rangle\langle\varphi_\alpha|$. This means the conditional state space lies inside a polytope with d vertices. It also means that if the conditional state space is too large to fit inside such a polytope, the state must be entangled. This point will be explored furtheron in the application to Werner states.

In [1] it is shown that the conditional state space of a pure bipartite state coincides with the complete state space on a n-dimensional matrix algebra, where n is the Schmidt number of the pure state. As the Schmidt number relates to the entanglement of the pure state, we again see a connection between the nature of the correlations and the properties of the conditional state space.

For general bipartite quantum states, the conditional state space can be modelled by a state space on an auxiliary system transformed by a completely positive map. We derive such a model in more detail for free Fermions.

2 Werner states

We first consider the example of qubit Werner states. A 2×2 Werner state on $\mathcal{M}_2 \otimes \mathcal{M}_2$ is a mixture between the uniform state $\frac{1}{4}\mathbb{1}_4$ and the maximally entangled state $|\Psi^-\rangle\langle\Psi^-|$, where $|\Psi^-\rangle = \frac{1}{\sqrt{2}}(|10\rangle - |01\rangle)$:

$$\rho_W = (1-\lambda)\frac{\mathbb{1}_4}{4} + \lambda|\Psi^-\rangle\langle\Psi^-|\,.$$

The state is separable for $0 \leq \lambda \leq \frac{1}{3}$ and entangled for $\frac{1}{3} < \lambda \leq 1$ [4]. We now consider the geometry of the conditional state space in function of λ.

The restriction $\mathrm{Tr}_1\rho_W$ of the Werner state to the first party is the maximally mixed state $\frac{1}{2}\mathbb{1}_2$. For the conditioning operator B, the requirement of unitality is then

$$\mathrm{Tr}\rho_W \mathbb{1}_2 \otimes B = \mathrm{Tr}_\mathcal{B}(\mathrm{Tr}_\mathcal{A}(\rho_W)B) = \mathrm{Tr}\frac{B}{2} = 1\,,$$

so $B = 2\rho'$ for some qubit state ρ'. The state conditioned on B becomes

$$\omega_B(A) = \mathrm{Tr}\rho_W A \otimes B = (1-\lambda)\mathrm{Tr}\frac{\mathbb{1}_2}{2}A + \lambda\langle\Psi^-|A \otimes 2\rho'|\Psi^-\rangle$$

If ρ' has a Bloch vector $\boldsymbol{r}$ then the functional $A \mapsto \langle \Psi^- | A \otimes 2\rho' | \Psi^- \rangle$ is a qubit state with Bloch vector $-\boldsymbol{r}$. Hence the possible conditional states are mixtures of a general qubit state with the uniform state $\frac{1}{2} \mathbb{1}_2$ with a weight λ. The conditional state space is thus a Bloch ball with radius λ.

For two qubits it is known that any separable state can always be decomposed as a mixture of at most 4 separable states [5]. This means that the conditional state space lies inside of a polygon with 4 vertices, i.e. a tetrahedron. Therefore, if the conditional state space can not be put inside a tetrahedron, the bipartite state must be entangled. This corresponds to what we see for the Werner state. The largest ball centered in the origin that fits inside of a tetrahedron has radius $1/3$. Hence, if the radius of the conditional state space, equal to the mixing parameter λ, becomes larger than $1/3$, the Werner state must be entangled.

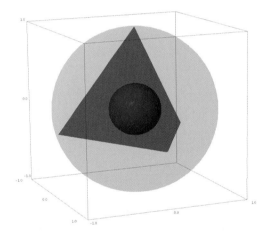

Fig. 1. The conditional state space of a Werner state with $\lambda = 1/3$ pictured within a tetrahedron and the Bloch ball.

3 Free fermions

The main goal of this paper is to present the characterization of conditional state spaces of free fermionic states. We first briefly introduce these states and what we mean by bipartite fermionic states, see [1] and references therein for more details.

The fermionic creation and annihilation operators obey the following relations:

$$\varphi \in \mathcal{H} \mapsto a^*(\varphi) \text{ is } \mathbb{C}\text{-linear,}$$
$$\{a(\varphi), a(\psi)\} = 0 \text{ and } \{a(\varphi), a^*(\psi)\} = \langle \varphi, \psi \rangle \mathbb{1}, \text{ (anticommutation)}$$

where $|\varphi\rangle$ and $|\psi\rangle$ are vectors of the one-particle or mode Hilbert space $\mathcal{H}$ of the system. The creation and annihilation operators generate the algebra $\mathcal{A}(\mathcal{H})$ of canonical anticommutation relations (CAR).

A free state ω_Q on $\mathcal{A}(\mathcal{H})$ is determined by a symbol which is a linear operator Q on $\mathcal{H}$ satisfying $0 \leq Q \leq \mathbb{1}$. The ω_Q-expectations of all monomials vanish except for those with an equal number of creation and annihilation operators (the gauge-invariant elements):

$$\omega_Q\big(a^*(\varphi_1)a^*(\varphi_2)\cdots a^*(\varphi_n)a(\psi_n)\cdots a(\psi_2)a(\psi_1)\big) = \det\Big(\big[\langle\psi_k,\, Q\,\varphi_\ell\rangle\big]\Big). \quad (1)$$

Positivity holds if and only if $0 \leq Q \leq \mathbb{1}$. The set of symbols

$$\mathcal{Q}(\mathcal{H}) = \{Q \mid Q \text{ bounded linear operator on } \mathcal{H} \text{ such that } 0 \leq Q \leq \mathbb{1}\} \quad (2)$$

is convex and weakly compact. Convexity at the level of symbols is very different from convexity at the level of the free states because (1) is non-linear in Q. Nevertheless it can be shown that a free state is pure, i.e. extreme in the full state space of $\mathcal{A}(\mathcal{H})$, if and only if its symbol is an orthogonal projector, i.e. an extreme point of $\mathcal{Q}(\mathcal{H})$.

Quantities like the entropy of free states are expressible in terms of symbols, e.g.

$$\mathsf{S}(Q) = -\mathrm{Tr}\,Q \log Q - \mathrm{Tr}(\mathbb{1} - Q) \log(\mathbb{1} - Q). \quad (3)$$

This formula is an example of the reduction of complexity in free fermionic states. Although ω_Q is a state on the CAR algebra $\mathcal{A}(\mathcal{H})$, its properties are fully determined by an operator on the one-particle space $\mathcal{H}$.

For distinguishable particles composition of subsystems amounts to tensoring their Hilbert spaces. For indistinguishable particles composition is done at the level of the one-particle space. The direct sum is the natural way of composing fermionic subsystems. Two free states on different CAR algebras $\mathcal{A}(\mathcal{H}_1)$ and $\mathcal{A}(\mathcal{H}_1)$ define a unique product state on $\mathcal{A}(\mathcal{H}_1 \oplus \mathcal{H}_2)$. If ω_i is an even state, i.e. a state vanishing on monomials of odd degree, on $\mathcal{A}(\mathcal{H}_i)$ for $i = 1, 2$, then there exists a unique state $\omega_1 \wedge \omega_2$ on $\mathcal{A}(\mathcal{H}_1 \oplus \mathcal{H}_2)$ such that

$$(\omega_1 \wedge \omega_2)(X_1 X_2) = \omega_1(X_1)\,\omega_2(X_2), \quad X_i \in \mathcal{A}(\mathcal{H}_i). \quad (4)$$

Here we consider more general states on composite CAR algebras.

Let P be an orthogonal projection on $\mathcal{H}$, then the restriction of the free state ω_Q on $\mathcal{A}(\mathcal{H})$ is a free state on the sub-CAR algebra $\mathcal{A}(P\mathcal{H})$ with symbol PQP. Conversely, an orthogonal decomposition $\mathcal{H} = \mathcal{H}_1 \oplus \mathcal{H}_2$ turns $\mathcal{A}(\mathcal{H})$ into a composite system with parties $\mathcal{A}(\mathcal{H}_i)$ up to a minor modification: $\mathcal{A}(\mathcal{H}_i)$ sits as a graded tensor factor in $\mathcal{A}(\mathcal{H})$ through the natural identification $a^*(\varphi_i) \mapsto a^*(\varphi_i \oplus 0)$. This is due to the fact that odd elements in $\mathcal{A}(\mathcal{H}_1 \oplus 0)$ anticommute with odd elements in $\mathcal{A}(0 \oplus \mathcal{H}_2)$. To simplify notation we shall often write $a^*(\varphi)$ instead of $a^*(\varphi \oplus 0)$.

4 Free fermionic conditional states

We now consider a free state on a bipartite fermionic system $\mathcal{A}(\mathcal{H}_1 \oplus \mathcal{H}_2)$ defined by a symbol Q with block matrix structure

$$Q = \begin{bmatrix} A & B \\ B^* & C \end{bmatrix}. \tag{5}$$

The aim is to characterize all free states on $\mathcal{A}(\mathcal{H}_1)$ that arise as conditional states of a bipartite free state. More precisely, to characterize

$$S_1^{\text{free}} = \Big\{ \omega_{\tilde{A}} \,\Big|\, \qquad \omega_{\tilde{A}} \text{ is a free state on } \mathcal{A}(\mathcal{H}_1) \text{ and}$$
$$\exists \text{ a gauge-invariant } Y \in \mathcal{A}(\mathcal{H}_2) \text{ such that}$$
$$Y \geq 0 \text{ and } \omega_{\tilde{A}}(X) = \omega_Q(XY), \ X \in \mathcal{A}(\mathcal{H}_1) \Big\}. \tag{6}$$

In [1], we obtained two characterizations of such conditional state spaces, one in terms of operator bounds on symbols, another using completely positive maps. We simply state the results here and refer to [1] for a full proof.

Proposition 1. *The weak*-closure of the set S_1^{free} of conditioned free states on $\mathcal{A}(\mathcal{H}_1)$ coincides with the set of free states on $\mathcal{A}(\mathcal{H}_1)$ whose symbols $\tilde{A}$ satisfy*

$$A - BC^{-1}B^* \leq \tilde{A} \leq A + B(\mathbb{1} - C)^{-1}B^*. \tag{7}$$

The second characterization uses an auxiliary state space and a completely positive map Γ that preserves the free nature of the states, i.e. mapping free states into free states.

Such a $\Gamma : \mathcal{A}(\mathcal{H}) \to \mathcal{A}(\mathcal{K})$ is determined by operators

$$R : \mathcal{H} \to \mathcal{K} \text{ and } S : \mathcal{H} \to \mathcal{H} \tag{8}$$

that satisfy

$$0 \leq S \leq \mathbb{1} - R^*R. \tag{9}$$

The action on a gauge-invariant monomial of order two is given by

$$\Gamma(a^*(\varphi)a(\psi)) = a^*(R\varphi)a(R\psi) + \langle \psi, S\varphi \rangle, \ \varphi, \psi \in \mathcal{H}. \tag{10}$$

This yields the following transformation of free states:

$$\omega_Q \circ \Gamma = \omega_{R^*QR+S}.$$

Proposition 2. *There exists a unique, free, minimal, identity preserving, completely positive map Γ from an auxiliary algebra $\mathcal{A}(\mathcal{H}')$ to $\mathcal{A}(\mathcal{H}_1)$ such that the weak*-closure of S_1^{free} is the pull-back of the free states by Γ. Stated differently, there exist operators R and S fulfilling $0 \leq S \leq \mathbb{1} - R^*R$ such that $\omega_{\tilde{A}}$ lies in the weak*-closure of S_1^{free} iff $\tilde{A} = R^*DR + S$ with $0 \leq D \leq \mathbb{1}$*

The validity of this proposition can be verified by taking

$$R = U \sqrt{B\, C^{-1} B^* + B\, (\mathbb{1} - C)^{-1} B^*} \text{ and } S = A - B\, C^{-1} B^*$$

and by checking that these operators fulfil $0 \leq S \leq \mathbb{1} - R^* R$ and that the set $\{R^* D R + S | 0 \leq D \leq \mathbb{1}\}$ coincides with the set of operators described in Proposition 1.

5 Examples

In the remainder of the text we explore some basic applications of the propositions.

Example 1. As a trivial example consider a product state, i.e. a state with $B = 0$. From either proposition it is clear that the conditional state space is zero-dimensional as it should be for a product state.

Example 2. Let's consider the case $\mathcal{H}_1 = \mathcal{H}_2 = \mathbb{C}$. The symbol of the bipartite free state then takes the form

$$Q = \begin{bmatrix} a & b \\ b^* & c \end{bmatrix},$$

with $a, b, c \in \mathbb{C}$. Positivity of the state ω_Q means $0 \leq Q \leq \mathbb{1}$, from which we see that $0 \leq a \leq 1$, $0 \leq a \leq 1$, $|b|^2 \leq ac$ and $|b|^2 \leq (1-a)(1-c)$.

Proposition 1 tells us that the conditional states are all $\omega_{\tilde{a}}$ with

$$a - \frac{|b|^2}{c} \leq \tilde{a} \leq a + \frac{|b|^2}{1-c}.$$

Note that the positivity requirements on a, b and c ensure that $0 \leq \tilde{a} \leq 1$. It is clear that if the symbols a and c are fixed, the correlations are determined by b and that the conditional state space increases accordingly with larger values of $|b|$.

It is also clear that by taking $R = e^{i\theta} \sqrt{\frac{|b|^2}{c} + \frac{|b|^2}{1-c}}$ and $S = a - \frac{|b|^2}{c}$ that this interval coincides with that prescribed by Proposition 2.

Example 3. The third example characterizes the conditional state space of pure free states. A quasi-free state is pure if and only if its symbol Q is a projector, i.e. $Q^2 = Q$. For a bipartite state with symbol

$$Q = \begin{bmatrix} A & B \\ B^* & C \end{bmatrix} \tag{11}$$

this means that

$$A^2 + BB^* = A,$$
$$B^*B + C^2 = C,$$
$$AB + BC = B, \text{ and}$$
$$B^*A + CB^* = B^*.$$

Through the polar decomposition of B, we get from the first equation that $B = \sqrt{A(\mathbb{1} - A)}U$ for some unitary U. Inserting this in the third equation, we get $C = U^*(\mathbb{1} - A)U$. From Proposition 1 now follows that $0 \leq \tilde{A} \leq \mathbb{1}$. The conditional state space becomes a full quasi-free state space, similar to what happens for pure states of distinguishable particles. The role that the Schmidt number played for pure distinguishable states is now played by the dimension of the image of A.

6 Conclusion

We introduced the concept of conditional state spaces for quantum systems, in particular for free Fermions. Possible applications of such spaces lie in finding quantum phase transitions in the sense of a quantum DLR equation [6] and in the description of quantum correlations such as entanglement. The latter was illustrated by an application to Werner states.

We also described two different characterizations of free fermionic conditional state spaces and explored their meaning in some examples.

References

1. M. Fannes, J. Wouters: J. Phys. A **42**, 465308 (2009).
2. O. Bratteli, D. Robinson: *Operator Algebras and Quantum Statistical Mechanics, Volume II*, 2nd edn (Springer, Berlin Heidelberg New York 2002).
3. M.Fannes, B. Nachtergaele, R.F. Werner: Commun. Math. Phys. **144**, 443 (1992).
4. R.F. Werner: Phys. Rev. A **40**, 4277 (1989).
5. A. Sanpera, R. Tarrach, G. Vidal: Phys. Rev. A **58**, 826 (1998).
6. M. Fannes, R.F. Werner: Helv. Phys. Acta **68**, 635 (1995).

Uncertainty related to position and momentum localization of a quantum state

Łukasz Rudnicki

Center for Theoretical Physics, Polish Academy of Sciences, Al. Lotników 32/46
02-668 Warszawa, Poland
rudnicki@cft.edu.pl

Abstract. This paper presents the uncertainty related to position and momentum localization of a quantum state in terms of entropic uncertainty relations. We slightly improve the inequality given in [I. Bialynicki-Birula, Phys. Rev. A **74**, 052101 (2006)], and introduce a new entropic measure with corresponding uncertainty relation.

1 Introduction

From the famous Heisenberg uncertainty principle we can learn that "it is impossible to prepare states in which position and momentum are simultaneously arbitrarily well localized" [1]. It means that a measurement of two observables dealing with position and momentum localization of a quantum state respectively shall expose some uncertainty. Our goal is to describe this uncertainty (originated from the complementarity between position and momentum variables) with the help of entropic uncertainty relations. To this end we shall use the Rényi entropy defined for a probability distribution $\{P_i\}$:

$$H_\alpha^{(P)} = \frac{1}{1-\alpha} \ln \left[\sum_i P_i^\alpha \right], \tag{1}$$

where $\alpha > 0$. In some cases we will also use the Shannon entropy defined by the following limit:

$$S^{(P)} = \lim_{\alpha \to 1} H_\alpha^{(P)} = -\sum_i P_i \ln P_i. \tag{2}$$

To assure the maximal simplicity of the formulas appearing in this paper we will restrict ourselves to a one-dimensional case, where the quantum state is described in the position representation by a normalized wave function $\psi(x)$. In the momentum representation the same state is described by the wave function $\tilde{\psi}(p)$ related to the previous one by the Fourier transformation:

$$\tilde{\psi}(p) = \frac{1}{\sqrt{2\pi\hbar}} \int_{\mathbb{R}} dx \, e^{-ipx/\hbar} \psi(x). \tag{3}$$

To explain how to describe the uncertainty in terms of entropic uncertainty relations let us consider two arbitrary observables (F and G) with corresponding hermitian operators $\hat{F}$ and $\hat{G}$. We assume that these operators have point spectrum and in general do not commute with each other. The eigenstates $\varphi_m(x)$ of the operator $\hat{F}$, and $\theta_n(x)$ of the operator $\hat{G}$ form two orthonormal bases:

$$\int_{\mathbb{R}} dx\, \varphi_m(x)\, \varphi_{m'}^*(x) = \delta_{mm'}, \qquad \int_{\mathbb{R}} dx\, \theta_n(x)\, \theta_{n'}^*(x) = \delta_{nn'}. \tag{4}$$

The following completeness relations are also satisfied:

$$\sum_m \varphi_m(x)\, \varphi_m^*(x') = \sum_n \theta_n(x)\, \theta_n^*(x') = \delta(x - x'). \tag{5}$$

With the observables F and G we can associate the probability distributions $\{|f_m|^2\}$ and $\{|g_n|^2\}$, where:

$$f_m = \int_{\mathbb{R}} dx\, \psi(x)\, \varphi_m^*(x), \tag{6}$$

$$g_n = \int_{\mathbb{R}} dx\, \psi(x)\, \theta_n^*(x). \tag{7}$$

The amplitudes f_m and g_n are connected by a unitary transformation U_{mn}:

$$f_m = \sum_n U_{mn} g_n, \quad g_n = \sum_m U_{mn}^* f_m, \quad \sum_n U_{mn} U_{m'n}^* = \delta_{mm'}, \tag{8}$$

where:

$$U_{mn} = \int_{\mathbb{R}} dx\, \varphi_m^*(x)\, \theta_n(x). \tag{9}$$

According to Riesz theorem [2, 3] we have the norm inequality:

$$\left[c_U \sum_m |f_m|^{2\alpha} \right]^{1/\alpha} \leq \left[c_U \sum_n |g_n|^{2\beta} \right]^{1/\beta}, \tag{10}$$

$$\frac{1}{\alpha} + \frac{1}{\beta} = 2, \quad \alpha \geq 1, \quad c_U = \sup_{(m,n)} |U_{mn}|. \tag{11}$$

By definition $0 \leq c_U \leq 1$. Taking the logarithm of both sides of the norm inequality (10) we obtain the uncertainty relation for the sum of the Rényi entropies:

$$H_\alpha^{(F)} + H_\beta^{(G)} \geq -2 \ln c_U. \tag{12}$$

For every given pair of observables (F, G) one can find the eigenstates φ_m and θ_n and calculate the value of c_U. Typically, when the commutator of $\hat{F}$ and $\hat{G}$ does not vanish, c_U shall be less than 1, what simply means that the values of the observables F and G cannot be simultaneously well determined. When this commutator is equal to 0 the operators $\hat{F}$ and $\hat{G}$ share common eigenvectors and $c_U = 1$.

2 Localized measurements

To include information about localization properties of the quantum state we shall change our previous set of observables. Let us divide both position and momentum space into equal bins. To describe this partition we introduce the characteristic function:

$$\chi_j(s) = \begin{cases} 1 & s \in \left[\left(j - \frac{1}{2}\right)\delta s, \ \left(j + \frac{1}{2}\right)\delta s\right] \\ 0 & \text{elsewhere} \end{cases}. \tag{13}$$

The index j labels the bins and δs denotes the bin's width. With the kth bin in the position space we associate an observable A_k with corresponding eigenstates $\varphi_{km}(x)$ forming in the line segment, described by $\chi_k(x)$, an orthonormal basis:

$$\int_{\mathbb{R}} dx\, \chi_k(x)\, \varphi_{km}(x)\, \varphi_{km'}^*(x) = \delta_{mm'}, \quad \sum_m \varphi_{km}(x)\, \varphi_{km}^*(x') = \delta_k(x - x'). \tag{14}$$

One should notice that the Dirac delta function in (5) differs from the one in (14) by its domain. The domain of $\delta_k(x - x')$ is restricted to the kth bin. Similarly in the lth bin in the momentum space, described by $\chi_l(p)$, we introduce an observable B_l with eigenstates $\theta_{ln}(p)$. We have:

$$\int_{\mathbb{R}} dp\, \chi_l(p)\, \theta_{ln}(p)\, \theta_{ln'}^*(p) = \delta_{nn'}, \quad \sum_n \theta_{ln}(p)\, \theta_{ln}^*(p') = \delta_l(p - p'). \tag{15}$$

For the observables A_k and B_l we have the probability distributions $\{|a_{km}|^2\}$ and $\{|b_{ln}|^2\}$, where:

$$a_{km} = \int_{\mathbb{R}} dx\, \chi_k(x)\, \psi(x)\, \varphi_{km}^*(x), \tag{16}$$

$$b_{ln} = \int_{\mathbb{R}} dp\, \chi_l(p)\, \tilde{\psi}(p)\, \theta_{ln}^*(p). \tag{17}$$

In this case the unitary transformation U_{kmln}:

$$a_{km} = \sum_{l,n} U_{kmln}\, b_{ln}, \quad b_{ln} = \sum_{k,m} U_{kmln}^*\, a_{km}, \tag{18}$$

reads:

$$U_{kmln} = \frac{1}{\sqrt{2\pi\hbar}} \int_{\mathbb{R}} dx\, \chi_k(x) \int_{\mathbb{R}} dp\, \chi_l(p)\, e^{ipx/\hbar}\, \varphi_{km}^*(x)\, \theta_{ln}(p). \tag{19}$$

Using the same arguments as before we obtain the uncertainty relation:

$$H_\alpha^{(A)} + H_\beta^{(B)} \geq -2\ln c_U, \quad c_U = \sup_{(k,m,l,n)} |U_{kmln}|, \tag{20}$$

where now the Rényi entropies are:

$$H_\alpha^{(A)} = \frac{1}{1-\alpha}\ln\left[\sum_{k,m=-\infty}^{\infty} |a_{km}|^{2\alpha}\right], \quad H_\beta^{(B)} = \frac{1}{1-\beta}\ln\left[\sum_{l,n=-\infty}^{\infty} |b_{ln}|^{2\beta}\right]. \tag{21}$$

The uncertainty relation (20) caries both information about the uncertainty of position and momentum localization, and uncertainties associated with the observables A_k and B_l. To calculate the uncertainty related only to the localization properties of the state we shall find the maximal value C_{max} among c_U's calculated for all possible choices of $\varphi_{km}(x)$ and $\theta_{ln}(p)$. To do this we will firstly use the Hölder inequality [4]:

$$|U_{kmln}| \leq \frac{1}{\sqrt{2\pi\hbar}} \left(\int_{\mathbb{R}} dx\, \chi_k(x) \, |\varphi_{km}(x)|^2 \right)^{1/2} \times \tag{22}$$

$$\left(\int_{\mathbb{R}} dx\, \chi_k(x) \left| \int_{\mathbb{R}} dp\, \chi_l(p)\, e^{ipx/\hbar} \theta_{ln}(p) \right|^2 \right)^{1/2}.$$

From the orthonormality condition (14) we obtain that the integral in the first parenthesis is equal to 1. Now we rewrite (22):

$$|U_{kmln}| \leq \sqrt{W[\theta_{ln}]}, \tag{23}$$

where we have introduced the functional $W[\theta_{ln}]$ of the form:

$$W[\theta_{ln}] = \int_{\mathbb{R}} dp\, \chi_l(p) \int_{\mathbb{R}} dq\, \chi_l(q)\, Q_k(p-q)\theta_{ln}(p)\, \theta_{ln}^*(q), \tag{24}$$

and the integral kernel $Q_k(p-q)$ is:

$$Q_k(p-q) = \frac{1}{2\pi\hbar} \int_{\mathbb{R}} dx\, \chi_k(x)\, e^{i(p-q)x/\hbar} = e^{-i\frac{k\delta x}{\hbar}(p-q)} \frac{\sin\left[\frac{\delta x}{2\hbar}(p-q) \right]}{\pi(p-q)}. \tag{25}$$

We can define a new function:

$$\Psi(p) = e^{-i\frac{k\delta x}{\hbar}(p+l\delta p)} \theta_{ln}(p+l\delta p), \tag{26}$$

in terms of which the functional W takes a simpler form:

$$W[\Psi] = \int_{\mathbb{R}} dp\, \chi_0(p) \int_{\mathbb{R}} dq\, \chi_0(q)\, Q_0(p-q)\Psi(p)\, \Psi^*(q). \tag{27}$$

We got rid of the indices k and l due to the translational symmetries both in coordinate and momentum spaces. To find the maximal possible value of the functional $W[\Psi]$ we have to solve the variational equation:

$$\frac{\delta}{\delta\Psi} \left(W[\Psi] - \lambda \int_{\mathbb{R}} dp\, \chi_0(p)\, |\Psi(p)|^2 \right) = 0, \tag{28}$$

where λ is the Lagrange multiplier associated with the normalization constraint. The equation (28) leads to the Fredholm integral equation of the second kind [5]:

$$\frac{1}{\pi} \int_{-1}^{1} ds\, \frac{\sin\left[\frac{\gamma}{4}(t-s) \right]}{t-s} \Psi_j(s) = \lambda_j \Psi_j(t), \tag{29}$$

where $\gamma = \delta x \delta p/\hbar$ is a dimensionless parameter which appears instead of the bin's widths δx and δp. The maximal value of $W[\Psi]$ is the largest eigenvalue λ_0 of (29) [1, 5]:

$$W[\Psi] \leq \lambda_0 = \frac{\gamma}{2\pi} \left[R_{00}\left(\frac{\gamma}{4}, 1 \right) \right]^2, \tag{30}$$

where $R_{00}(s, t)$ is one of the radial prolate spheroidal wave functions of the first kind[1] [6]. Thus, with the help of the variational method we have found that:

$$C_{max} = \sqrt{\lambda_0} = \sqrt{\frac{\gamma}{2\pi}} R_{00}\left(\frac{\gamma}{4}, 1\right). \tag{31}$$

For every finite value of the parameter γ we have $C_{max} < 1$ which is the signature of uncertainty of position and momentum localization. The corresponding entropic uncertainty relation is the Maassen-Uffink type inequality [3, 7]:

$$H_\alpha^{(A)} + H_\beta^{(B)} \geq -2\ln C_{max}. \tag{32}$$

3 Probability distributions for localization in coordinate and momentum space

The probability distributions $\{|a_{km}|^2\}$ and $\{|b_{ln}|^2\}$ carry both information about localization and information about the observables A_k and B_l. In order to obtain a pure information about localization we shall trace out the observables' degrees of freedom:

$$q_k = \sum_m |a_{km}|^2 = \int_{\mathbb{R}} dx\, \chi_k(x)\, |\psi(x)|^2, \tag{33}$$

$$p_l = \sum_n |b_{ln}|^2 = \int_{\mathbb{R}} dp\, \chi_l(p)\, \left|\tilde{\psi}(p)\right|^2. \tag{34}$$

The q_k coefficient has an interpretation of the probability of finding the particle in the k'th bin in the coordinate space. The p_l coefficient can be interpreted in a similar manner. The probability distributions q_k and p_l are affected by the Heisenberg uncertainty principle, thus we have [1]:

$$\forall_{k,l}\quad q_k + p_l \leq 1 + \sqrt{\lambda_0}, \tag{35}$$

where λ_0 was defined in (30). The way of finding the relation (35) leads to the same integral equation (29). From (35) we can easily find that:

$$\forall_{k,l}\quad q_k p_l \leq \frac{1}{4}\left(1 + \sqrt{\lambda_0}\right)^2. \tag{36}$$

The inequality (36) is suitable to be used together with the Shannon entropies, because the sum of the Shannon entropies is:

$$S^{(q)} + S^{(p)} = -\sum_{k,l} q_k p_l \ln q_k p_l, \tag{37}$$

and we are able to find a simple lower bound:

$$S^{(q)} + S^{(p)} \geq -\sum_{k,l} q_k p_l \ln\left[\frac{1}{4}\left(1 + \sqrt{\lambda_0}\right)^2\right] \geq -2\ln\left[\frac{1}{2}\left(1 + C_{max}\right)\right]. \tag{38}$$

We can call it the Deutsch type inequality [7, 8].

---

[1] in the Wolfram Mathematica's notation it is $\texttt{SpheroidalS1}[0, 0, \mathsf{s}, \mathsf{t}]$.

In spite of the fact that $H_\alpha^{(A)} \geq H_\alpha^{(q)}$ and $H_\beta^{(B)} \geq H_\beta^{(p)}$ there is another lower bound for the sum $H_\alpha^{(q)} + H_\beta^{(p)}$ which in the quantum regime ($\gamma < 1$) is significantly better than (32). The probability distributions (33, 34) fulfill the chain of inequalities:

$$
\begin{aligned}
\left(\sum_{k=-\infty}^{\infty} q_k^\alpha \right)^{1/\alpha} &\leq (\delta x)^{1-1/\alpha} \left(\int_\mathbb{R} dx\, |\psi(x)|^{2\alpha} \right)^{1/\alpha} \\
&\leq (\delta x)^{(1-1/\alpha)} \left(\frac{\alpha}{\pi} \right)^{-1/2\alpha} \left(\frac{\beta}{\pi} \right)^{1/2\beta} \left(\int_\mathbb{R} dp\, \left| \tilde\psi(p) \right|^{2\beta} \right)^{1/\beta} \qquad (39) \\
&\leq (\delta x)^{(1-1/\alpha)} (\delta p)^{(1/\beta-1)} \left(\frac{\alpha}{\pi} \right)^{-1/2\alpha} \left(\frac{\beta}{\pi} \right)^{1/2\beta} \left(\sum_{l=-\infty}^{\infty} p_l^\beta \right)^{1/\beta},
\end{aligned}
$$

where the first and the last one are the Jensen inequalities and in the middle there is the famous Beckner inequality [9]. From (39) one can derive the uncertainty relation [10]:

$$
H_\alpha^{(q)} + H_\beta^{(p)} \geq -\frac{1}{2} \left(\frac{\ln \alpha}{1-\alpha} + \frac{\ln \beta}{1-\beta} \right) + \ln \pi - \ln \gamma. \qquad (40)
$$

This lower bound is better than (32) and (38) for small values of the parameter γ, but unfortunately it becomes negative (and in fact meaningless) for:

$$
\gamma \geq \frac{\pi}{\alpha} (2\alpha - 1)^{\frac{2\alpha-1}{2(\alpha-1)}}. \qquad (41)
$$

In the case of the Shannon entropies ($\alpha \to 1$) the condition (41) reads $\gamma \geq e\pi$.

4 Summary

The best calculated lower bounds for the sum of Shannon entropies which can be treated as a measure of uncertainty of position and momentum localization of the quantum state are:

$$
S^{(A)} + S^{(B)} \geq -\ln \left[\min \left\{ \frac{\gamma}{e\pi}, \frac{\gamma}{2\pi} \left[R_{00} \left(\frac{\gamma}{4}, 1 \right) \right]^2 \right\} \right], \qquad (42)
$$

in the case of the probability distributions $|a_{km}|^2$ and $|b_{ln}|^2$, and:

$$
S^{(q)} + S^{(p)} \geq -\ln \left[\min \left\{ \frac{\gamma}{e\pi}, \frac{1}{4} \left[1 + \sqrt{\frac{\gamma}{2\pi}} R_{00} \left(\frac{\gamma}{4}, 1 \right) \right]^2 \right\} \right], \qquad (43)
$$

in the case of the probability distributions (33, 34) which completely describe the localization properties.

Acknowledgement. I would like to thank Iwo Bialynicki-Birula who was the supervisor of this work. This research was supported by the grant from the Polish Ministry of Science and Higher Education.

References

1. P. Busch, T. Heinonen, P. Lahti: Phys Rep. **452**, 155 (2007).
2. M. Riesz: Acta Math. **49**, 465 (1927).
3. H. Maassen, J.B.M. Uffink: Phys. Rev. Lett. **60**, 1103 (1988).
4. D.S. Mitrinović: *Analytic Inequalities*. Springer-Verlag, Heidelberg (1970).
5. T. Schürmann: Act. Phys. Pol. B **39**, 587 (2008).
6. M. Abramowitz, I.A. Stegun: *Handbook of Mathematical Functions*. Dover, New York, (1964).
7. I. Bialynicki-Birula,L. Rudnicki: *Uncertainty relations related to the Rényi entropy*, in *Statistical Complexity - Applications in Electronic Structure*, ed. Sen, K.D. (2011). Preprint arXiv:1001.4668.
8. D. Deutsch: Phys. Rev. Lett. **50**, 631 (1983).
9. W. Beckner: Ann. Math. **102**, 159 (1975).
10. I. Bialynicki-Birula: Phys. Rev. A **74**, 052101 (2006).

Wave comeback? On classical representation of quantum averages

Andrei Khrennikov

International Center for Mathematical Modelling in Physics and Cognitive Sciences, Linnaeus University, Växjö, S-35195, Sweden
`Andrei.Khrennikov@lnu.se`

Abstract. We present fundamentals of a prequantum model with hidden variables of the classical field type. In some sense this is the comeback of classical wave mechanics. Our approach also can be considered as incorporation of quantum mechanics into classical signal theory. All quantum averages (including correlations of entangled systems) can be represented as classical signal averages and correlations.

1 Introduction

This conference collected a strong group of young active researchers who will definitely make in future important contributions to the theory of quantum and classical statistical correlations. Therefore, it is very important to send this group the following message: in spite of all no-go theorems, quantum randomness can be reduced to randomness of classical fields fluctuating on a "prequantum time scale". The latter is very fine compared with the presently approachable scale of measurements. We are not able to approach it right now. However, in future this time scale will be approached and it will be possible to monitor directly fluctuating prequantum random field. Thus, the analogy between quantum and classical (random) signals is much closer than it is commonly believed.

We present fundamentals of our model, the *prequantum classical statistical field theory* (PCSFT) [1]– [12]. This is a model with hidden variables of the classical field type. In some sense, this is the comeback of classical wave mechanics to the quantum domain. Our approach also can be considered as incorporation of quantum mechanics into classical signal theory. All quantum averages (including correlations of entangled systems) can be represented as classical signal averages and correlations.

2 Classical fields as hidden variables

Main message: *Quantum randomness is reducible to randomness of classical fields.*

PCSFT unifies two of Einstein's dreams: to reduce quantum randomness to classical randomness, and to create a pure wave model of physical reality. Classical fields are selected as the hidden variables. Mathematically, these are functions ϕ : $\mathbf{R}^3 \to \mathbf{C}$ (or more generally $\to \mathbf{C}^k$) which are square integrable, i.e., elements of the L_2-space. The latter condition is standard in classical signal theory; in particular, for the electromagnetic field this is just the condition that the energy is finite:

$$\int_{\mathbf{R}^3} (E^2(x) + B^2(x))dx = \int_{\mathbf{R}^3} |\phi(x)|^2 < \infty, \tag{1}$$

where $\phi(x) = E(x) + iB(x)$ is the Riemann-Silbertstein vector. Thus, the state space of our prequantum model is $H = L_2(\mathbf{R}^3)$. Formally, the same space is used in QM, but we prefer to emphasize coupling with the classical signal theory. For example, the quantum wave function should satisfy the normalization condition $\int_{\mathbf{R}^3} |\phi(x)|^2 = 1$, but a PCSFT-state can be any vector of H. These prequantum waves evolve in accordance with Schrödinger's equation; formally, the only difference is that the initial condition ϕ_0 is not normalized by 1. Thus these PCSFT-waves are closely related to Schrödinger's quantum waves. However, opposite to Schrödinger and the orthodox Copenhagen interpretation, the wave function of the QM-formalism is not the state of a quantum system. In complete accordance with Einstein's dream, it is associated with an ensemble. However, this is not an ensemble of quantum systems. This is an ensemble of classical fields or it is better to say a *classical random field*, a random signal.

A random field (at fixed instant of time) is a function $\phi(x, \omega)$, where ω is the random parameter. Thus, for each ω_0, we obtain the classical field, $x \mapsto \phi(x, \omega_0)$. Another picture of the random field is the H-valued random variable, where each ω_0 determines a vector $\phi(\omega) \in H$. A random field is given by a probability distribution on H. To simplify considerations, we can consider a finite-dimensional Hilbert space, instead of $L_2(\mathbf{R}^3)$ (as people often do in quantum information theory). In this case, our story is on H-valued random vectors, where $H = \mathbf{C}^n$. This is the ensemble model of the random field. In the rigorous mathematical framework, it is based on *Kolmogorov probability space*: $(\Omega, \mathcal{F}, \mathbf{P})$ where Ω is a set, $\mathcal{F}$ is a σ-algebra of its subsets, and $\mathbf{P}$ is a probability measure on $\mathcal{F}$.

However, as is well known from classical signal theory, one can move from the ensemble description of randomness to the time series description – under the *ergodicity hypothesis*.

3 Covariance operator interpretation of wave function

Main message: *The wave function is not a field of probabilities, neither a physical field. It encodes the covariance operator of a prequantum random field.*

In our model, the wave function ψ of the QM-formalism encodes the prequantum random field: $\phi \equiv \phi_\psi$. The QM-terminology, "a quantum system in the state ψ", is translated into the PCSFT-terminology, a "random field." In my model, the ψ-*function determines the covariance operator of the prequantum random field*.

For simplicity, in this introductory section we consider the case of a single, i.e., noncomposite, system, e.g., an electron (nonrelativistic, since the present PCSFT is nonrelativistic theory[1]), and we neglect for the moment (again for simplicity) the fluctuations of the vacuum.

In this situation, the covariance operator (normalized by dispersion) is given by the orthogonal projector on the vector ψ :

$$D_\psi = \psi \otimes \psi, \qquad (2)$$

i.e., $D_\psi u = \langle u, \psi \rangle \psi, \ u \in H.$

We also suppose that that all *prequantum fields have zero average:* $E\langle y, \phi \rangle = 0, y \in H$, where E denotes the classical *mathematical expectation* (average, mean value). By applying a linear functional y to the random vector ϕ, we obtain a scalar random variable. In the L_2-case, we get a family of scalar random variables: $\omega \mapsto \xi_y(\omega) \equiv \int y(x)\phi(x, \omega)dx, y \in L_2$. We recall that the covariance operator D of a random field (with zero average) $\phi \equiv \phi(x, \omega)$ is defined by its bilinear form: $\langle Du, v \rangle = E\langle u, \phi \rangle \langle \phi, v \rangle, u, v \in H.$

Under the additional assumption that the prequantum random fields are *Gaussian*, the covariance operator uniquely determines the field. Although this assumption seems to be quite natural both from mathematical and physical viewpoints, we should be very careful. In the case of a single system, we try to proceed as far as possible without this assumption. However, the PCSFT-description of composite systems is based on Gaussian random fields, see [8]–[12] for details. If $H = \mathbf{C}^n$, where $\phi(\omega) = (\phi_1(\omega), ..., \phi_n(\omega))$, then the zero average condition is reduced to $E\phi_i = 0, i = 1, 2$, and the covariance matrix $D = (d_{kl})$, where $d_{kl} = E\phi_k\bar{\phi}_l$.

The random field $\phi(x, \omega)$ corresponding to a pure quantum state is not L_2-normalized. Its L_2-norm $\|\phi\|^2(\omega) \equiv \int_{\mathbf{R}^3} |\phi(x, \omega)|^2 dx$ fluctuates depending on the random parameter ω.

4 Quantum observables from quadratic forms of the prequantum field

Main message: *In spite of all no-go theorems (e.g., Kochen-Specker), a natural functional representation of quantum observables exists.*

In PCSFT quantum observables are represented by corresponding quadratic forms of the prequantum field. A self-adjoint operator $\widehat{A}$ is considered as the symbolic representation of the PCSFT-variable:

$$\phi \mapsto f_A(\phi) = \langle \widehat{A}\phi, \phi \rangle. \qquad (3)$$

We remark that f_A can be considered as a function on the phase space of classical fields: $f_A \equiv f_A(q, p)$, where $\phi(x) = q(x) + ip(x)$.

The average of this quadratic form with respect to the random field determined by the wave function ψ coincides with the corresponding quantum average:

$$\langle f_A \rangle = \langle \widehat{A}\psi, \psi \rangle. \qquad (4)$$

---

[1] It seems that there are no barriers to develop a relativistic variant of PCSFT. I plan to do this in future.

Here

$$\langle f_A \rangle = E f_A(\phi) = \int_H f_A(\phi) d\mu_\psi(\phi)$$

is the classical average and μ_ψ is the probability distribution of the prequantum random field $\phi \equiv \phi_\psi$ determined by the pure quantum state ψ. In the real physical case, H is infinite-dimensional; the classical average is given by the integral over all possible classical fields; probabilistic weights of the fields are determined by ψ. Thus, the quantum formula for the average of an observable was demystified: $\langle \widehat{A} \rangle_\psi \equiv \langle \widehat{A}\psi, \psi \rangle = \int_H f_A(\phi) d\mu_\psi(\phi)$. It can be obtained via the classical average procedure.

5 Quantum and prequantum interpretations of Schrödinger's equation

Main message: *Schrödinger's equation does not describe the dynamics of a wave of probability. The same equation plays a double role. On the one hand, it describes the dynamics of a physical random field. On the other hand, it encodes the dynamics of the covariance operator of this field.*

Before going to the PCSFT-dynamics, we consider the Schrödinger equation in the standard QM-formalism:

$$ih\frac{\partial \psi}{\partial t}(t,x) = \widehat{\mathcal{H}}\psi(t,x), \tag{5}$$

$$\psi(t_0, x) = \psi_0(x), \tag{6}$$

where $\widehat{\mathcal{H}}$ is Hamiltonian, the energy observable. Although Schrödinger tried to interpret $\psi(t,x)$ as a classical field (e.g., the electron field), the conventional interpretation is the probabilistic one, due to Max Born.

We recall that a time dependent random field $\phi(t,x,\omega)$ is called a *stochastic process* (with the state space H). The dynamics of the prequantum random field is described by the simplest stochastic process which is given by *deterministic dynamics with random initial conditions*.

In PCSFT, the Schrödinger equation, but with random initial condition, describes the dynamics of the prequantum random field, i.e., the prequantum stochastic process can be obtained from the same mathematical equation as it was used in QM for the dynamics of the wave function:

$$ih\frac{\partial \phi}{\partial t}(t,x,\omega) = \widehat{\mathcal{H}}\phi(t,x,\omega), \tag{7}$$

$$\phi(t_0, x, \omega) = \phi_0(x, \omega), \tag{8}$$

where the initial random field $\phi_0(x,\omega)$ is determined by the quantum pure state ψ_0. The standard QM provides the knowledge of the covariance operator of this random field.

Roughly speaking we combined Schrödinger's and Born's interpretations: the ψ-function of QM is not a physical field, but, for each t, it determines a random physical field – the H-valued stochastic process $\phi(t,x,\omega)$.

The PCSFT dynamics (7), (8) matches with the standard QM-dynamics (5), (6) – by taking into account the PCSFT-interpretation of the wave-function, see (2). Denote by $\rho(t)$ the covariance operator of the random field $\phi(t) \equiv \phi(t, x, \omega)$, the solution of (7), (8). Then $\rho(t) \equiv \rho_\psi(t) = \psi(t) \otimes \psi(t)$, where $\psi(t)$ is a solution of (5), (6).

Such a simple description can be used only for a single system by ignoring fluctuations of the vacuum. In the general case – for a composite system, e.g., a biphoton system, in the presence of vacuum fluctuations – the Schrödinger dynamics of the ψ-function encodes only the dynamics of the covariance operator of the prequantum stochastic process, see [8]–[12] for details. The situation is essentially more complicated than in the case of a single system. We found [8]–[12] that it is possible to construct a few different prequantum dynamics which match (on the level of correlations) with the QM-dynamics (given by the Schrödinger equation for the wave function), (5), (6).

6 Towards prequantum determinism?

Main message: *The background field (vacuum fluctuations) is everywhere. This is the space by itself that fluctuates permanently. It seems that these random fluctuations are inescapable.*

From the PCSF-viewpoint, the source of quantum randomness is the randomness of initial conditions (if one neglects vacuum fluctuations) – due to the impossibility to prepare a non-random initial prequantum field $\phi_0(x)$.

It may be that in future very stable and precise preparation procedures will be elaborated. The output of such a procedure will be a deterministic field $\phi(x)$, i.e., the random fluctuations will be eliminated. For such prequantum states, quantum randomness of measurements (the latter also should be very stable and sensitive) will be eliminated.

However, it may be that such a dream of creation of supersensitive "subquantum" technologies will never come true. In such a case PCSFT will play the role of classical statistical mechanics of prequantum fields. Unfortunately, there are a few signs that it might really happen. First of all, it might be that the scale of prequantum fluctuations is very fine, e.g., the Planck scale. In such a case, it would be really impossible to prepare a deterministic prequantum field. Another reason is that the PCSFT-model presented up to now was elaborated for noncomposite quantum systems, e.g., a single electron. Extension of PCSFT to composite systems, e.g., a pair of entangled photons or electrons is based on a more complicated model of prequantum randomness, see [8]–[12] for the detailed presentation. We should complete the model by fluctuations of the background field (zero point field, vacuum fluctuations), in the same way as in SED. In reality, the latter fluctuations are always present. Therefore, Einstein's dream on determinism is obscured by the presence of the background field. If this field is irreducible, in the sense that it is the fundamental feature of space, then deterministic prequantum fields will be never created.[2]

---

[2] As I understood from conversations with Gerard't Hooft, it is his viewpoint: quantum randomness can be reduced to classical, but determinism cannot be restored, because of fluctuations of the vacuum, see also [13].

If this background field is simply a noise which can be (at least in principle) eliminated, then we can dream of the creation of deterministic prequantum fields. Nevertheless, even if vacuum fluctuations are really irreducible, some prequantum field effects (which are not described by QM) might be observed, at least in principle, see [9].

7 Random fields corresponding to mixed states

Main message: *A density matrix is the normalized covariance operator of a prequantum random field.*

We now consider the general quantum state given by a density operator ρ. (We still work with noncomposite quantum systems.) By PCSFT it determines the covariance operator of the corresponding prequantum field (under normalization by its dispersion):

$$D_\rho = \rho. \tag{9}$$

The dynamics of the prequantum field $\phi(t, x, \omega)$ is also described by the Schrödinger equation, see (7), (8), with the random initial condition $\phi_0(x, \omega)$. The initial random field has the probability distribution μ_{ρ_0} having zero mean value and the covariance operator

$$D(t_0) = \rho_0.$$

Under the assumption that all prequantum random fields (at least corresponding to so called "quantum systems") are Gaussian, the initial probability distribution is determined in a unique way. In the general (non-Gaussian) case, we loose the solid ground. The $\phi_0(x, \omega)$ can be selected in various ways - it can be any distribution having the covariance $D(t_0)$. We could not exclude such a possibility. It would simply mean that macroscopic preparation procedures are not able to control even the probability distribution (only its covariance operator).

Denote by $\rho(t)$ the covariance operator of the random field $\phi(t) \equiv \phi(t, x, \omega)$ given by (7), (8) with ϕ_0 having the covariance operator $\rho(t_0) = \rho_0$. Then $\rho(t)$ satisfies the von Neumann equation. However, $\rho(t)$ has a classical probability interpretation as the covariance operator.

8 Background field

Main message: *QM is a formalism of measurement with calibrated detectors.*

As was mentioned, randomness of initial conditions has to be completed by taking into account vacuum fluctuations. In our model, the background field (vacuum fluctuations) is of the white noise type. It is a Gaussian random field with zero average and covariance operator

$$D_{\text{background}} = \varepsilon I, \ \varepsilon > 0.$$

It is a stationary field, so its distribution does not change with time.

The presence of this field is a source of additional mathematical complexity of the PCSFT model. It is well known that the probability distribution of white noise is not σ-additive on $H = L_2$. Roughly speaking, it is concentrated on a "larger space", some space of generalized functions, say $H_- = \mathcal{S}(\mathbf{R}^3)$, the space of Schwartz distributions. To escape this difficulty, we can proceed in the finite-dimensional Hilbert space H.

Consider (by using the QM-language) a quantum system in the mixed state ρ_0. It determines the prequantum random field $\phi_0 \equiv \phi_0(x, \omega)$ with the covariance operator

$$\tilde{D}(t_0) = \rho_0 + \epsilon I.$$

The value of $\varepsilon > 0$ is not determined by PCSFT, but it could not be too small by a purely mathematical reason, see [8]–[12]. Now consider the solution $\phi(t)$ of the Schrödinger equation (7), (8) with initial condition ϕ_0. Its covariance operator can be easily found:

$$\tilde{D}(t) = D(t) + \varepsilon I,$$

where $D(t)$ is the covariance operator of the process in the absence of the background field, $D(t) = \rho(t)$, where $\rho(t)$ satisfies the QM-equation for evolution of the density operator, the von Neumann equation. Thus, on the level of the dynamics of the covariance operator, the contribution of the background field is very simple. However, on the level of the field dynamics, the presence of vacuum fluctuations crucially changes the field behavior.

Consider the prequantum random field $\phi_0(x, \omega)$ corresponding to a pure quantum state ψ_0. Now (in the presence of the background field) the prequantum random field $\phi_0(x, \omega)$ is not concentrated on the one dimensional subspace[3]

$$H_{\psi_0} = \{\phi = c\psi_0 : c \in \mathbf{C}\};$$

the vacuum fluctuations smash it over H.

In canonical QM, the background field of the white noise type is neglected. And it is the right strategy for a formalism describing measurements on the random background. However, in an ontic model, i.e., a model of reality as it is, this background field should be taken into account. Its neglection induces a rather mystical picture of quantum randomness.

We shall see that in the PCSFT-formalism, the background field plays the fundamental role in the derivation of Heisenberg's uncertainty relation, see [10]. Roughly speaking, Heisenberg's uncertainty is a consequence of vacuum fluctuations. The presence of these fluctuations is a real barrier for the physical realization of Einstein's dream of deterministic prequantum evolution.

Mathematically, we can consider the prequantum field $\phi_0 = \phi_0(x)$ which does not depend on a random parameter; starting with ϕ_0, we solve the Schrödinger equation (7). In reality, it means that we should be able to isolate ϕ_0 and $\phi(t)$ from the background field. It is not clear whether it will be possible even in future. Thus, although "quantum randomness" is reducible to classical randomness (of initial conditions and vacuum fluctuations), it might be "inescapable."

I propose, in fact, the comeback to classical field theory; roughly speaking, in the spirit of early Schrödinger: Maxwell's classical field theory should be extended

---

[3] In the absence of vacuum fluctuations the covariance operator of the random field $\phi_\psi(x, \omega)$ corresponding to a pure state ψ is given by the orthogonal projector on ψ, see (2); the corresponding Gaussian measure is concentrated on one dimensional subspace based on ψ.

to "matter waves", in the spirit of Einstein and Infeld [14]. Of course, this comeback is not the dream of the majority of those who nowadays are not afraid to speculate on prequantum models. Nevertheless, I do not think that PCSFT is too cheap a completion of standard QM. At least, I hope that, in contrast to Bohmian mechanics, Einstein might accept PCSFT as one of the possible ways beyond QM. In any event, the Laplacian mechanistic determinism was totally excluded from PCSFT; reality became blurred (in the sense of Schrödinger), but still reality.

8.1 Coupling with classical signal theory

Main message: *QM is a version of classical signal theory. It is about very noisy and (temporally and spatially) singular signals.*

In the previous sections, we used the ensemble approach to random fields. However, it is possible to switch from the ensemble representation of randomness of signals (as ensembles of classical fields) to the time representation of randomness (as signals fluctuating on a fine time scale). This can be done on the basis of the *ergodic hypothesis.*

Instead of a random field $\phi(x, \omega)$, which is distributed with some probability distribution $d\mu(\phi)$ on H, we consider a time dependent signal $\phi(s) \equiv \phi(s, x)$, where $x \in \mathbf{R}^3, s \in [0, +\infty)$. Then, for each functional $f(\phi)$ such that $\int_H |f(\phi)| d\mu(\phi) < \infty$, we have (under the assumption of ergodicity):

$$\langle f \rangle_\mu \equiv \int_H f(\phi) d\mu(\phi) = \lim_{T \to \infty} \frac{1}{T} \int_0^T f(\phi(s)) ds \equiv \langle f \rangle_\phi. \tag{10}$$

Consider two time scales: τ is a fine scale and $T >> \tau$ is a rough time scale. In QM, the latter is the scale of measurements and Schrödinger's dynamics, and τ is the scale of fluctuations of the prequantum field. Thus,

$$\langle f \rangle_\phi \approx \frac{1}{T} \int_0^T f(\phi(s)) ds, \tag{11}$$

where s denotes the time variable at the τ-scale. We call the T-scale the *quantum (experimental) time scale*[4].

For each signal $\phi(s, x), x \in \mathbf{R}^3$, PCSFT determines time averages of physical variables; for example, for the energy variable of the prequantum electromagnetic field, the "photonic field", we have:

$$\langle f \rangle_\mu \approx \frac{1}{T} \int_0^T \left(\int_{\mathbf{R}^3} (E^2(s, x) + B^2(s, x)) dx \right) ds. \tag{12}$$

In time representation, the covariance operator (its bilinear form) of the signal $\phi(s, x)$ is given by

$$\langle Du, v \rangle = \lim_{T \to \infty} \frac{1}{T} \int_0^T \left(\int_{\mathbf{R}^3} u(x) \overline{\phi(s, x)} dx \int_{\mathbf{R}^3} \phi(s, x) \overline{v(x)} dx \right) ds$$

---

[4] This terminology does not match with the modern tradition in cosmology and superstring theory to refer to the quantum scale as the Planck scale. However, our terminology matches well with Bohr-Heisenberg view on QM as theory of measurements performed by macroscopic devices.

$$\approx \frac{1}{T} \int_0^T \left(\int_{\mathbf{R}^3} u(x)\overline{\phi(s,x)}dx \int_{\mathbf{R}^3} \phi(s,x)\overline{v(x)}dx \right) ds, \qquad (13)$$

where $u(x), v(x)$ are two test signals, $u, v \in L_2(\mathbf{R}^3)$.

This paper was completed during my visit to University of Rome-2 (September 2010) I would like to thank Prof. Luigi Accardi for fruitful discussions on quantum foundations and hospitality.

References

1. A. Khrennikov: J. Phys. A: Math. Gen. **38**, 9051 (2005).
2. A. Khrennikov: Found. Phys. Letters **18**, 637 (2005).
3. A. Khrennikov: Physics Lett.A **357**, 171 (2006).
4. A. Khrennikov: Found. Phys. Lett. **19**, 299 (2006).
5. A. Khrennikov: Nuovo Cimento B **121**, 505 (2006).
6. A. Khrennikov: Physics Lett. A **372**, 6588 (2008).
7. A. Khrennikov: J. Russian Laser Research **30**, 472 (2009).
8. A. Khrennikov: Europhysics Lett. **88**, 40005 (2009).
9. A. Khrennikov: Physica E **42**, 287 (2010).
10. A. Khrennikov: Phys. Scrip. **81**, 065001 (2010).
11. A. Khrennikov: J. of Russian Laser Research **31**, 191 (2010).
12. A. Khrennikov: Found. Phys. **40**, 1051 (2010).
13. G. 't Hooft: Quantum Mechanics and Determinism. Preprint arxiv:hep-th/0105105.
14. A. Einstein and L. Infeld: *Evolution of Physics: The Growth of Ideas from Early Concepts to Relativity and Quanta* (Simon and Schuster, New York 1961)